Orcas Island Library District

WITHDRAWN

P9-AGW-377

3 3628 00041 9084

THE CONTRARY FARMER

THE REAL GOODS INDEPENDENT LIVING BOOKS

Paul Gipe, *Wind Power for Home & Business: Renewable Energy for the 1990s and Beyond*

Michael Potts, *The Independent Home: Living Well with Power from the Sun, Wind, and Water*

Gene Logsdon, *The Contrary Farmer*

Real Goods Solar Living Sourcebook: The Complete Guide to Renewable Energy Technologies, Eighth Edition, edited by John Schaeffer

Real Goods Trading Company in Ukiah, California, was founded in 1978 to make available new tools for helping people live self-sufficiently and sustainably. Through seasonal catalogs, a quarterly newspaper (*The Real Goods News*), the *Solar Living Sourcebook*, as well as a book catalog and retail outlets, Real Goods provides a broad range of renewable-energy and resource-efficient products for independent living.

"Knowledge is our most important product" is the Real Goods motto. To further its mission, Real Goods has joined with Chelsea Green Publishing Company to co-create and co-publish the Real Goods Independent Living Book series. The titles in this series are written by pioneering individuals who have firsthand experience in using innovative technology to live lightly on the planet. Chelsea Green books are both practical and inspirational, and enlarge our view of what is possible as we enter the next millennium.

Ian Baldwin, Jr.
President, Chelsea Green

John Schaeffer
President, Real Goods

Orcas Island Library District

GENE LOGSDON

THE CONTRARY FARMER

CHELSEA GREEN PUBLISHING COMPANY
POST MILLS, VERMONT

Designed by Kate Mueller, Electric Dragon Productions

© 1993 by Gene Logsdon. All rights reserved.
No part of this book may be transmitted in any form by any
means without permission in writing from the publisher.
Printed in the United States of America.

1 2 3 4 5 6 7 8 9 10

Chelsea Green Publishing Company
P.O. Box 130, Route 113
Post Mills, Vermont 05058

"The Contrariness of the Mad Farmer," by Wendell Berry, is
from *Farming: A Hand Book*, A Harvest/HBJ Book, Harcourt
Brace Jovanovich, copyright © 1967, 1968, 1969, 1970.
Reprinted with permission.

Library of Congress Cataloging-in-Publication Data

Logsdon, Gene.
The contrary farmer / Gene Logsdon.
p. cm.
Includes bibliographical references (p.) and index.
ISBN 0–930031–67–9
1. Agriculture. 2. Far life. 3. Logsdon, Gene. 4. Agriculture—
United States. 5. Farm life—United States. I. Title.
S523.L67 1994
630'.1—dc20   94–2659

Contents

This book is dedicated to my friend, Dave Smith, on whose contrary farm the marauding wild boars understand retribution but the old redwoods grow peacefully, knowing they will not be cut down as long as he lives.

The Contrariness of the Mad Farmer

I am done with apologies. If contrariness is my
inheritance and destiny, so be it. If it is my mission
to go in at exits and come out at entrances, so be it.
I have planted by the stars in defiance of the experts,
and tilled somewhat by incantation and by singing,
and reaped, as I knew, by luck and Heaven's favor,
in spite of the best advice. If I have been caught
so often laughing at funerals, that was because
I knew the dead were already slipping away,
preparing for a comeback, and can I help it?
And if at weddings I have gritted and gnashed
my teeth, it was because I knew where the bridegroom
had sunk his manhood, and knew it would not
be resurrected by a piece of cake. "Dance" they told me
and I stood still, and while they stood
quiet in line at the gate of the Kingdom, I danced.
"Pray" they said, and I laughed, covering myself
in the earth's brightnesses, and then stole off gray
into the midst of a revel, and prayed like an orphan.

When they said "I know that my Redeemer liveth,"
I told them "He's dead." And when they told me
"God is dead," I answered "He goes fishing every day
in the Kentucky River. I see Him often."
When they asked me would I like to contribute
I said no, and when they had collected
more than they needed, I gave them as much as I had.
When they asked me to join them I wouldn't
and then went off by myself and did more
than they would have asked. "Well, then" they said
"go and organize the International Brotherhood
of Contraries," I said "Did you finish killing
everybody who was against peace?" So be it.
Going against men, I have heard at times a deep harmony
thrumming in the mixture, and when they ask me what
I say I don't know. It is not the only or the easiest
way to come to the truth. It is one way.

Wendell Berry
from Farming: A Hand Book

Gratitudes

My thanks go first to Dave Smith, to whom this book is dedicated. Now an executive with Real Goods Trading Corporation and formerly the co-founder of Smith and Hawkens, Dave staked me to the writing of this book both with money and with an indefatigable flow of encouragement. Those who think that American business has grown decadent in the pursuit of self-aggrandizement need to know Dave as well as I have come to know him.

Secondly, a special thanks to Jim Schley, my editor at Chelsea Green Publishing Company. I have been edited by some very astute people, but never one with such a passion for the details of the English language and such a gimlet eye not only for my violations of its rules but for the rhythm of my sentences and the cultural rhyme of my occasional outbursts of criticism and sarcasm. Gawd. There were times when his skill and dedication sent me nearly clawing up the oak tree outside my window—mostly because he was mostly right and I was mostly wrong. If this manuscript reads well to you, the credit should go entirely to Jim's tireless efforts. In fact, so pervasive is his hand in every chapter that I am going to depart from the usual writerly custom and say that any mistakes left are entirely Jim's. Just kidding.

Obviously, for the Contrary Farmer to succeed, he must have a contrary wife and family and a most contrary bunch of relatives and friends. You all know who you are and how much you've helped. And how I can never repay any of you adequately.

I give special thanks to two friends and university teachers, Kamyar Enshayan and David Orr. They are as contrary in academia as I am in agriculture, willing to point out problems in their university establishments with great courage and little regard for their own career security. They are true Teachers, and have inspired me greatly to keep pressing on. If only we had colleges full of people like them, then colleges might make sense again.

The Ramparts People

I remember clearly the day when I was twelve, hunting morel mushrooms with my father, when I informed him excitedly that I had decided to take my dog and my rifle and go deep into the wilderness to live. I would build a cabin on a mountainside by a clear running stream, and live out my days happily on broiled trout, fried mushrooms, and hickory nut pie. I would achieve advanced degrees in the art of living, bestowed on me by Nature, and I would know many things not even Einstein or my stupid schoolteacher dreamed of.

I thought that he would approve, since he was forever retreating to the solitude of woods and river bank and farm field himself. But he almost frowned, suggesting gently in a voice that sounded as if he were saying what he thought he was supposed to say, not what he really felt, that I needed to be thinking about making my way in the world and contributing something to it.

Unfortunately I tried to follow his advice and it took me until I was forty-two to realize that I knew what was better for me when I was twelve. And having hunted everywhere for the peculiar kind of freedom I had tried to articulate that day, I came back to my boyhood home—the place of my beginnings—and found it. What I learned in the process was to follow my own mind because worldly wisdom invariably springs from notions that are largely erroneous. The only really good advice that holds up in all situations is: Always make friends with the cook.

For a while, I thought Americans had lost the desire for independence—the kind of independence that defines success in terms of how much food, clothing, shelter, and contentment I could produce for myself rather than how much I could buy; the kind of freedom that examines the meaning of life, not the meaning of cholesterol; the kind of freedom that allows me to say what I think in public without fear that my words will be "bad for business," the fear that keeps my rich acquaintances in town in silent bondage, trading their freedom of speech for dollars. (Not a one of them will publicly say what they privately believe: that President Clinton is as mad as ex-President Bush for dropping "well-intentioned" bombs on defenseless countries, and so the polls all appear to approve an act of outright terrorism.)

Then I started hearing about other people who were even more independent than I dared to dream: people deliberately removing themselves from the protection of the great god, Grid, because only beyond the blessings of the holy public utility could they find affordable land of their own: and also people, excluded from even that kind of frontier, who were turning ghettos into edenic gardens. I became acquainted with a university music professor who farmed with horses and in retirement manufactured modern horsedrawn machinery; a scientist who discovered that composted sewage sludge protects vegetable plants from disease; a man who homesteaded with his family in an isolated rural area to start a million-dollar business creating beautiful and useful items out of waste wood even while a rare disease slowly incapacitated his muscular coordination; a Vietnamese immigrant who figured out how to use duckweed (green pond scum) to purify wastewater and then made a nutritious protein supplement out of the scum; a rock star who bought a thousand-acre farm and turned it back into a wilderness that produces more food than the farm did; a Pulitzer Prize-winning journalist who quit his career to become an organic market gardener; a famous cartoonist who built a sewage system for his huge office complex that uses the shit and urine from his fifty workers to grow exotic plants, fish, and mussels, and then discharges pure water back into the environment; a contractor who uses scrap tires, earth, and beer cans to build houses that run entirely on the sun.

The voice of the turtle can be heard again, ringing through the land, as the old Wyandots and Mohegans who once roamed my farm would say—a new surge of creative energy that moves the earth in a direction of self-redemption and sustainability that not the richest PAC nor the oldest institutionalized claptrap can stop.

We are pioneers, seeking a new kind of religious and economic freedom. We flee the evils that centralized power always generates. Our God does not reside in the inner sanctums of cathedrals, but walks with us, hoeing in the fields. Sometimes I see Him checking the bluebird houses for murderous starlings and house sparrows and give Him hell for inventing the nest-robbing bandits. He smiles and reminds me that stupid scientists brought the starling and house sparrow to America, not Him.

We are circumspect about our economic institutions. We do not bank on paper money within marble walls, but invest in sun and soil and sweat and the tools that make sweat more productive.

I think of us as the Ramparts People. In all ages we have camped on the edges of the earth, the buffer between our more conventional and timid brethren and those nether regions where, as the medieval maps instructed, "there be dragons and wild beestes." It is our destiny to draw the dragon's fire while the mainstream culture hides behind its disintegrating deficit and damns us for shattering its complacency. So be it.

The hickory nut pie is excellent.

CHAPTER 1

At Ease with the
Work of Farming

*. . . all of us will come back again to hoe in the ground . . .
or hand-adze a beam, or skin a pole, or scrape a hive—
we're never going to get away from that. We've been living a
dream that we're going to get away from it . . . Put that out
of our minds . . . That work is always going to be there.*

Gary Snyder, in *The Real Work, Interviews and Talks, 1964–1979*

My Uncle George liked to say that lazy farmers built the best fences because they didn't want to do the work over in a few years. That was his way of saying that successful contrary farming depends crucially on reducing manual labor to a minimum by skill instead of expensive machines and making the hard work that remains more enjoyable. This is particularly important for those of us who must combine farm work with another job or career to make a living.

This ability to manage manual labor efficiently requires a list of attitudes and skills as long as a hoe handle, but might be summed up by the scene of my grandfather, Henry Rall, grinning mischievously as he drove his horses while sitting on the rocking chair he had wired to the harrow. He even offered a reason to so pamper himself: the extra weight made the harrow do a better job of leveling the soil. Grandpaw Rall was exhibiting the most necessary skill to enjoying hard work: technological cleverness. Grandpaw Logsdon was good at that too. He pounded a stake into the middle of his large, grassed barnyard, and attached one end of a length of rope to the stake and the other to his lawnmower out on the edge of the lawn. Sure enough, the mower would run by itself in an ever decreasing circle as the rope wound around the stake, mowing most of the grass while Grandpaw cackled and drank hard cider in the shade.

Just as important as technological cleverness is what I call handiness. Good athletes are gifted with handiness—excellent agility and coordination combined with an inborn sense about how to apply muscular power at just the right moment, location, and thrust to gain the most effective power. Society praises as science and art the ability to swing a ball bat, golf club, or tennis racket skillfully. If we would expend a fraction of that kind of attention and honor on the hoe, axe, shovel, and pitchfork, we might be surprised about how much work that humans could accomplish without help of fossil fuel-gulping machines. They might not even realize that they were performing what journalists who don't farm call "back-breaking" work. Is there any more back-breaking work than playing middle linebacker? Glorifying work by making contests of it is precisely how society in pre-industrial civilization made physical work more or less enjoyable. One of my boyhood heroes, farmer Noble Goodman, was one of Ohio's great softball pitchers and also the Ohio state champion cornhusker in 1937.

It is difficult to generalize about hard work, because its definition depends on who's talking. Another of the farmers who influenced me when I was young, Henry Bils, was an immigrant from Belgium who during the first half of this century worked his way to farming success despite overwhelming obstacles. He thought little of weeding and thinning four acres of sugar beets a day with a hand hoe, sixty acres a season. Most farmers would consider his approach to work excessive. But not Henry. He was working for himself, and that made all the difference. "He had *vision*," his grandson, also Henry, says today. "He liked to work hard because in his mind he could see that it would pay off." It sure did. One year in the 1940s he made $15,000 from his hand-hoed sugar beets, a lot of money then and enough to pay off his debts in one grand slam. We can't do that sort of thing in farming today because we have been stupid enough to sneer the hoe into near oblivion.

Henry did not flee the farm when things got tough, unlike the Hamlin Garlands of the literary world, who ever after wrote condescending and denigrating books about "back-breaking drudgery" and economic failure. If Henry had written a book about farming (which he would have considered unendurably painful work), he would have told how exciting and rewarding his life was, even after he had bought a poor farm and gone broke on it. He would have pointed out his mistake and then told how he tried again on good land and succeeded beyond his dreams. Nor did his sons rebel against the hard work he submitted them to, as the Hamlin Garland school of fiction liked to claim. The sons all went on to be successful farmers too. Where love is at work, work is mostly play.

My parents worked very hard at farming when they had to, but they had a genius for making games out of work. Picking up ears of corn knocked off the stalks by the binder was a dull early winter chore in the old days, but became for us an exciting hunt for arrowheads as we walked across the bare ground of the corn stubble. Between arrowheads, Mom would recount stories endlessly from the books she was reading or the latest movie she had seen. Or she and Dad would get in an argument about religion, which was even more entertaining.

Both my parents and my wife's parents extolled hard work, insisting that the era of real horsepower was just as much fun and far less stressful than the high-tech days of later years. When the work did get grindingly physical, as in hay or grain harvests, a farmer could afford to hire help, they pointed out. Or neighbors got together and made a party of the work. "The young people went from farm to farm on winter evenings to husk the corn from the bundles of stalks stored in the barns," Dad would recall his father recalling. "Anyone husking out a red ear—and there were lots of red ears in the days before hybrid corn—got to kiss his or her sweetheart."

That story would prompt my wife's mother, Helen Downs, to recall: "We were just as proud of our buggies as young people today are of their cars, and we didn't have to spend all our lives paying for them either. It was more fun courting in a buggy too, because you could let the horses do the driving." Then she would pause, smile, and add: "When we went to town we took food along to sell. *The grocer always owed us money.*"

The truth is that farming at its worst is no more physically punishing than operating a restaurant, brokering commodities on the floor of the Chicago Board of Trade, or training for the Olympics. Yet our culture glamorizes these stressful occupations and clings to its image of farming as drudgery despite all evidence to the contrary. My brother-in-law, Morrison Downs, likes to say: "I left the farm for the city so I wouldn't have to work weekends. Now I am working weekends while the daggone farmers are all off fishing."

The real shame is that so few people know enough about the biological world of agriculture to appreciate how fascinating the work actually can be. A farmer of deep ecological sensitivity is to the plow jockey on his 200-horsepower tractor what a French chef is to the legions of hamburger handlers at fast food chains. The chef's work is infused with artistic, scientific, and spiritual satisfactions; the hamburger handler's is infused only with the ticking of a time clock. To the plow jockey, soil is a boring landscape of clods that need to be crushed. To the ecological farmer, every clod holds a wondrously exotic, tropical-like world of brilliantly colored microorganisms, the very stuff of life.

Nevertheless, there is much work associated with even a small cottage farm like our thirty-two acres. Making that work enjoyable is a kind of calling, I think. Not everyone is cut out for it, although I am sure that there are thousands of people going through life dissatisfied (I was one of them for a while) because they do not know that they were born to be nurturers—farmers. Sometimes, as a compromise, they become gardeners, and that's okay too.

This calling, by which physical work can be rendered enjoyable and interesting (surely more so than jogging) requires certain characteristics that may be learned, but that I believe are mostly inborn. The first is a love of home. People with a true vocation to contrary farming find so much fascination in the near-at-hand that they feel no need to wander the world in search of truth, or beauty, or amusement. Like the great naturalist Henri Fabre, who turned his backyard into a lifelong, living laboratory for the study of insects, true farmers see their farms and their communities as a source of never-ending discovery, a microcosm of the world. They see the grand canyons and tropical rain forests, the city lights fantastic, the now much-trodden wildernesses, the history of civilization ebbing and flowing, all repeated in their own neighborhoods and villages. If they wish to heighten their awareness of how the outer world is reflected in their lives, they can "travel" the world by book, or by radio, television, telephone, and computer. They learn that people are the same everywhere and that the way to enjoy humanity (or at least learn to endure its absurdities) is to cultivate the people and places of their own community. One can dine as well in a country cottage in Ohio on standing rib roast and homemade apple pie as in gay Paree on chateaubriand and crème brulée. More to the point, we can enjoy chateaubriand and crème brulée in a country cottage in Ohio. With this sensibility, a farmer avoids the attitude that most often makes farm work burdensome: he knows he is not missing something grand and great down the road someplace.

Right here in our neighborhood there are dramatic stories waiting to be written about nearly every farm and village home I have been privileged to enter.

In a little country cemetery close by, there is a grave of a woman who became pregnant, not with the man to whom she was betrothed, but to her father's hired man. Her husband-to-be renounced her. She committed suicide. He spit in her coffin while the carpenter was making it. That all happened many years ago, but someone still places a bouquet of flowers by her grave every year in May and nobody seems to know who does it.

Just north of the cemetery is the place where legend says the Bower boys beat my great-grandfather nearly to death and left him lying in the road because he was an immigrant farmer from Germany, and, horrors, a Catholic to boot. The family that lived in the house in view of the beating (their descendants, the Hollansheads, live there still) nursed great-grandfather back to health. He told them that he would some day own the Bowers' farm, and he did.

At the very next house, a windsock rippled in the breeze above the barn roof for more than thirty years, though there neither was, nor is, an airport for miles around. Why? Walter and Berenice Kail operated a dairy farm there for many years, and their son, John, always wanted to be a pilot. He finally succeeded and went to Korea with the Navy. Walter kept a long, level swath of grass in the field behind the barn for a landing strip so that when John came home from the war to take over the farm, he could handily continue his flying. But John was killed in action, shot down in his plane. A few months later, when the parents where still sick with grief, a bad storm swept over the farm, after which

5

a young man appeared at their door, grinning. "I just landed my plane on your field," he said. "The storm forced me down. I don't know what I would have done without that landing strip." The young man was Bill Dyviniak, a photographer for the *St. Louis Dispatch*. When he heard the explanation for the landing strip, he was profoundly affected. He became, in a way, the devoted son the Kails had lost, returning to the farm year after year, to this day, though Walter is gone now and Berenice lives in a retirement home. Until the farm was sold, Bill Dyviniak kept a windsock flying on the barn peak, a memorial to what he and the Kail family believe was a miraculous occurrence. By any measure, it was.

Where can I experience the world any more deeply? I am reminded of Andrew Wyeth and the Kuerner farm next to his home in Pennsylvania, about which I have written often and will have more to say in a practical vein later in this book. Wyeth has done hundreds of paintings on that farm, including many of his most famous works. Has he exhausted the possibilities of this humble, unpretentious-looking bit of farmland? By no means. Karl Kuerner, the third Karl Kuerner, farmborn and farm-raised and himself an artist of renown who also paints the home farm continuously (on the jacket of this book is one of his paintings), told me recently that in discussing the artistic possibilities of the place and its people, Wyeth remarked to him: "Karl, we haven't even hit the tip of this iceberg yet."

This ability to see extraordinary beauty and drama in a farm landscape is shared by real farmers, and is another reason that the work remains endurable if not enjoyable even in the most trying situations. The geometry of fields and garden plots never ceases to please the landlover's eye, even when sweat blurs the vision. There is, for example, a constant change of colors in the landscape as the sun moves over it. A field of wheat can turn into a rippling crimson lake at sunset. Tree trunks that were conventionally brown in the morning turn astonishingly orange in the dying light of certain magical evenings, especially in winter. Then in the moonlight, the trunks turn pitch black, with a contrast so sharp against the snow that it can take your breath away if the cold air does not.

The contrary farmer also enjoys hard work out of a sort of muleheaded stoicism. I like getting hot, tired, and dirty putting up hay because it feels so good to clean up in the evening, sit on the porch, and sip lemonade, especially if it is spiked with gin.

Beyond these psychological aids to ease the work of farming, there are the actual skills and methods that can make the work easier. The first rule is not to do anything nature will do for you. The new emphasis on all-season grazing of livestock with rotated pastures (again, I'll say much more, later) is an excellent example: why make all that hay and haul it to the barn if the cows can graze it off themselves, even through at least part of winter?

But the principle of letting nature do the work is far more complex, involving what Swedish scientist Stefin Delin, way ahead of his time, observed a few years ago: "It may well be that the biological processes are many magnitudes of order more efficient than the industrial ones." This idea suggests fields of knowledge not much explored yet and terms not yet coined that some day will describe how diversity in nature can lessen hard work in farming, not with bigger machines, but with ecological cleverness. As a way to get at this intriguing but still hazy idea, walk with me in spirit over our little farm where biological diversity is our first order of business. On this farm lives a human family along with several families of sheep, chickens, cows, bees, hogs, and in my more naive days, riding horses. Nurturing all of us and being nurtured by us are families of corn, oats, wheat, orchard trees, grasses, legumes, berries, and garden vegetables, the whole domestic tribe living in a sort of hostile harmony with the wild food chain: animals, insects, and plants in such diversity that I have not been able to name them all. On our little farm, I have identified 130 species of birds, 40 species of wild animals (not counting coonhunters), over 50 species of wildflowers, at least 45 tree species, a myriad of gorgeous butterflies, moths, spiders, beetles, etc., and about 593,455,780 weeds.

What does that diversity have to do with easing work? Does that not in fact sound like a guarantee for increasing work? Actually, not. For example, in the chemicalized fields that surround our farm, Canada thistle is a most noxious weed. The thistledown blows into our fields too, and the seeds sometime take root. The expert way to get rid of Canada thistle is to spray them (work) with a suitable herbicide (expense, meaning you have to work to pay for it), even though it is obvious all around us that herbicides do not control the thistle very well. But we don't spray the Canada thistle and it is not a problem for us. The reason why not is bound up in the inextricable webs of diverse life on our farm. First of all, Canada thistles prefer tilled soils, where they behave like the early Christians. The more of them you behead, the more they multiply. Since most of our farm is in thick permanent or semi-permanent sods where the thistle seeds are not as apt to establish themselves, we have an advantage in controlling them. Secondly, when the thistles do come up in the pasture, our sheep will nibble on them despite the prickles, thus impairing the thistles growth. Also two different bugs, the three-lined plant bug and the other I can't identify, riddle the thistle leaves until the infected plants stop growing. Most interestingly, a disease often strikes the thistles growing in sod, turning the tops white before they blossom and eventually killing them. Along with a pasture mowing or two, which I have to do anyway, Canada thistles are effectively controlled without one lick from me in the way of extra work.

I believe that the more diversity of species on a farm, the more the various forms of life keep each other from achieving out-of-balance population relative to the other species. This increasing diversity means more than merely "balance," which is a negative accomplishment. Increasing diversity means to me increasing biological dynamism which leads to an increasing amount of total food produced without increasing the amount of human labor or purchased agricultural supplies. The most obvious example is growing clover. Clover works with rhizobia bacteria in the soil to draw nitrogen from the air and make it available to itself and other subsequent plants without any effort or cost to me. A factory to extract nitrogen from the air costs millions of dollars and society's tendency is then to use the nitrate so produced to make gunpowder, not to enrich soil.

Ten acres of our thirty-two lie about a mile from the main farm. This is old growth forest which we bought to save it from the bulldozer. We manage it for an important crop: wood for fuel to keep us warm in winter, wood for construction lumber, and wood for our son's woodworking business. We have to work to turn that wood into fuel and lumber, but nature does all the work of producing it. What would be the energy efficiency of humans producing steel compared to nature producing wood?

Our home twenty acres are divided into six parts. Two acres along the road hold our vegetable gardens and orchard and woodworking shop. Then come about five acres of woodland in which the house and barn are nestled. The trees act as a windbreak to protect the barn in winter from cold westerly winds so that the environment around the buildings is fairly calm even on the most blizzardy days. The winter work of caring for the animals is thus made pleasant. The protection from the trees also means we need less fuel for the house in winter.

Behind this woods are about twelve acres of open land, divided by fences into plots of permanent pasture, temporary pasture, and grain crops. In one of the pasture plots, we dug a pond to avoid the work of hauling water to the plot that would otherwise be necessary in rotational grazing. Behind this area, at the rear of the property are another two acres of wooded pasture, through which the creek meanders. Our farm thus contains all the kinds of farm and wildlife habitat in north-central Ohio: garden, lawn, orchard, woodland, grassland, cultivated ground, creek, pond, and wetland. This panoply of habitat and the abundance of food it produces means that we are literally besieged by deer, raccoons, rabbits, squirrels, groundhogs, opossums (they make a nest in my grain harvester in the barn every winter), chipmunks, geese, ducks, and coyotes. We could easily obtain from this quasi-wilderness our yearly supply of meat if we so chose, and avoid the work of raising domestic animals altogether. I would guess that our farm sustains a yield of about two bushels of groundhogs per acre as a by-product of ecological farming. Young groundhog is not bad eating either.

As all these life forms interact with each other, they create effects that individually they are incapable of. For example, cow flaps draw earthworms to dine on the organic matter. Young trees that have crept into the meadow over the years from the adjoining woodlot draw the cows to their shade. The cow-manure-earthworm-tree environment draws woodcocks to the farm. These birds come for the earthworms under the cow flaps and under the moist dirt bared by tree shade and cow hooves. Not incidentally, the combination has also produced on occasion a fairy ring of edible mushrooms. And also not incidentally, the animal manure is all the while being broken down and returned to enrich the earth. All we have to do is stand and watch in awe and pick the mushrooms.

9

Sometimes wild animals work quite directly for us, like hired hands. My honeybees pollinate our crops and then provide us with honey. I know farmers who still let hogs and beef cattle harvest their corn. Mike Reicherts, a well-known Iowa farmer, says his hogs have learned how to knock the stalks over to get to the ears. "It is really something to see," he says with a grin, "a hog walk up to a stalk, look up at the ear of corn on it, and deliver a tremendous blow with the side of its snout and wham! Down comes the stalk. Somebody ought to get that on a camcorder."

My sheep clean out fencerows for me with their grazing and also save me much mowing. When a sheep dies, the buzzards soon swoop in out of the blue and gorge themselves on the carcass, as loathesome and gluttonous a sight as I have seen on our farm, but for that reason, fascinating too. The point is that the buzzards perform a useful activity from my point of view (not to mention the buzzard's) by saving me the job of burial.

Although we appear to live in a very tame, intensively farmed area, hardly a week goes by what we do not experience some unusual or unexpected little adventure which lightens and even makes gladsome the work. A pale green luna moth fluttering in the porch light; a fungus that looks like a little pile of sand; another that looks strikingly like a human penis; an ant "milking" its herd of aphids; a killdeer nest right in the middle of our gravel driveway. And three years after we planted pawpaw trees, the gorgeous zebra swallowtail butterfly, which feeds only on paw-paw, landed daintily on the tractor.

Learning to let nature do work for you applies to gardening and landscaping too. Many people mow their lawns twice as often as they need to, to the detriment of the lawn, and before long they are complaining that their place is too big to keep up.

Often, the rules that landscapists lay down for trimming trees (or foresters, for thinning a young stand of hardwood trees) instruct you to undertake work that, if you wait a year or two, will be done by natural shading. Working hard at building and turning compost heaps makes some gardeners happy and results in a wonderful soil amendment, but you can save lots of energy by just spreading leaf and grass waste as mulch and letting it rot to compost—in place, in its own good time.

If diversity is the first major worksaver, the second is timeliness. For example, it is crucially important to control weeds when they are tiny and easy to destroy. After weeds grow even two inches tall, controlling them becomes unpleasant work.

Timeliness can be practiced another way: by not biting off more than you can chaw, as the old saying goes. Almost all beginning gardeners who love their work plant gardens that are too large and then don't have time to tend them properly. Good French intensive gardeners can raise more on a hundred square feet than I am presently raising on three times that much space because they can concentrate water, soil nutrients, and their labor on a smaller area. And although I expend on thirty acres the same amount of time that a large operator spends, with several workers, on a thousand acres, my costs will be lower because my payroll is zero and my tools much cheaper, while my production *per acre* is much higher. I can focus all my skills and time on comparatively few acres. This economic verity becomes even truer as the number of acres farmed diminishes below twenty.

"In the United States, we've always talked about the fact that as farms become larger they become more efficient," Hugh Popenoe at the University of Florida has observed. "But we're talking about comparing a fifty-acre farm to a five-thousand-acre farm. We've never talked about farms of two, three, or four acres. As farms become smaller than three acres, yields start increasing dramatically."

No matter how small the farm, easing the work is better achieved if there are many activities in progress, spread over the entire year so that at no one time does work become overwhelming. Briefly, here's how spreading the work load works on our farm.

In January we have little to do other than the usual daily chores of feeding the animals, keeping them in clean straw, and feeding the stove in our living room with wood. We spend a lot of time reading and watching television and making big plans we never put into action.

In February, as the snow (if any) melts, I get into the woodcutting mode. If cold winds blow from the west, I cut on the east side of the woods, and vice versa. Out of the wind, the winter woods is more pleasant than the summer woods. At the end of the month, we tap a few maple trees and boil down a little maple sap.

In March, we shear the sheep and butcher two hogs. Butchering is distasteful work to me, but family and friends join in and with only two hogs, the job becomes almost trivial. I broadcast clover seed on the dormant wheat and sometimes on pastures. Woodcutting continues. I try to build several hundred feet of new fence every March and make any repairs needed in the existing fencing. Installing new fencing is work, but not nearly as hard as trying to keep animals inside deteriorating old fences.

April is lambing time. I turn the animals on pasture about mid-month or sooner. Toward the end of the month, as soon as the soil is dry enough, I disk the corn stubble field and plant it to oats. We also walk over the pastures, hoeing out any burdock that might later produce burrs to tangle in the sheeps' wool.

The work in May reaches high tide no matter how carefully I have spread the load. The priority job is getting garden vegetables and field corn planted and then continuously cultivating until July so that weeds never become a problem. Asparagus is in full production. But there will always be time on those first warm, bugless days to shed clothes and enjoy the new sun in the sweet spring air. Birdwatching combines well with sun-bathing. Watch especially for the Sharp-Eyed Prude that flits about on angel wings keeping the world safe for the clothing industry. You can usually identify it by its song: an irritable *tsk tsk tsk*.

June means haymaking, my hardest work. But in the chapter on that subject, you will see how I have considerably lessened the time and labor of that job with "technological cleverness." Also we dare not slack up on weed cultivation, except to go fishing. Fish in farm ponds bite best in June. I add another super (honey compartment) to each of the beehives in June. And eat more strawberries than I should.

July is the third and last of the big labor months, though I have not missed a softball game yet. This is the month we butcher the broilers bought in May as chicks. The second cutting of hay comes in now, the wheat and oats are harvested, and the straw barned for winter bedding. Time to rotate livestock to a different pasture and clip the one they have been grazing with tractor and mower. Lambs may need worming. Pick raspberries and blackberries. I swim in the pond with eyes closed, pretending I'm on the Riviera, wherever that is.

August is slow-down time on the farm, but putting food by in the kitchen is at its peak. Tomatoes, sweetcorn, stringbeans, peppers, peaches, plums, and so on must all be canned or frozen for winter's fare. We do not try to do all of a crop in one freezing or canning bout, but do small amounts several times. The work is less grueling that way. Also now is the time I usually haul manure out of the barn and give the pastures a second clipping.

Corn harvest begins in September. We feed the sweet corn stalks, relieved of their ears, to the sheep and cows, and if pastures have dried up, cut some green field corn for them too. As the field corn matures, I begin to husk it from the stalks, or in some cases, cut stalks with ears intact and set them up in shocks. Toward the end of the month, I disk the old oat stubble and plant wheat.

In October we finish the corn husking and remove the top compartments of the beehives for our share of the honey crop. We gather hickory nuts and walnuts and put the cider press into action. We make apple cider, apple pie, apple vinegar, applesauce, apple everything and still, in most years, there are plenty for the sheep and cows.

In November, the most important farm job is getting the old hay field plowed. Usually this is the month we saw logs from our woods into lumber. Firewood cutting begins now, along with making bittersweet and wild grapevine wreaths for the coming holidays. We sell the lambs, butcher our beef calf, and put a ram in with the ewes.

On nice days in December, we cut wood. There is much feverish activity in the woodworking shop where we are busy making Christmas gifts. Otherwise there is time to write letters to faraway friends, and snooze by the warm woodstove.

There is a daily rhythm to the work, especially where farm animals are involved. Morning and evening chores are inevitable with animals, feeding and watering them, gathering eggs, and milking the cow if a calf is not nursing. But the time involved in this daily routine with our forty sheep (more or less), a cow and calf, two hogs, and thirty chickens is seldom more than an hour a day, and in summer when everything is out on pasture, hardly any time at all. Even if I doubled the number of animals, I have now learned enough skills and shortcuts that the chore time involved would only increase a little.

To get a weekend away from home, we pick times of the year when the animals are on pasture, or provide them with enough food inside to last for two days. We have made arrangements with other contrary farmers to do each other's necessary chores if emergencies arise and we have to be away for longer than two days.

There's another consideration about manual labor in farming that is becoming more important as time progresses. Modern society is losing the knowledge of how to do anything in a direct, hands-on, manually-skilled kind of way. I don't know if the words "progressive" or "advanced" can be applied to a nation that can no longer function without certain technologies over which individuals have no control. The more local communities become dependent on centralized powers for food, clothing, and shelter, the more they become enslaved to that power. The typical American farmer today spends more time wrangling over subsidies at his local Agricultural Stabilization and Conservation Service office than he does planting corn.

I like what Jay Dorsey, a young agricultural engineer at Ohio State University, has written in a memo to his fellow engineers, protesting the loss of practical knowledge in farming. He says, in part:

> The highly technological agriculture we prescribe today has adversely affected the management skills and ability of the farmer. In the move to gain more control, the practitioner is losing touch with some basic agricultural principles that are effective in any agricultural system, regardless of the level of technology used, such as: stability through diversification; timing tillage and planting for weed and insect control instead of getting it all done as early as possible; and the use of rotations for free nitrogen, soil building, and weed and insect control. Farmers have lost much of the "feel" (practical knowledge) for how to farm because modern technologies have trivialized that knowledge. As a result, farmers don't "know" their farms or their soils as well as they used to, and in a sense they have lost the flexibility that would allow them to adjust to adverse environmental conditions. Farmers (and other groups of course) who used to be considered artisans are becoming little more than technicians.

As an example, there are very few professional farmers who know how to milk a cow by hand anymore, and I daresay none with the muscle tone to milk five in a row as my mother and I each used to do when Dad was busy in the fields. Even more alarming, the typical dairy cow today has been genetically re-shaped to sport teats that are fine for milking machines but too small to be hand-milked.

If farmers are becoming ignorant of practical knowledge, think how much more so society at large suffers from this illiteracy. Most people don't know how to feed, clothe, or shelter themselves, much less build a house or repair a roof. Hardly anyone, including the so-called trained mechanics, know how to fix the ailments of electronically-controlled cars. The mechanics don't fix these cars; they merely keep replacing old parts with new ones until the vehicle runs again.

But the best way I can describe the dilemma we are in is to tell a true story. A young, well-educated woman I know, with her heart very much in the right place, decided to grow a garden last year. She planted lots of potatoes. They grew wonderfully. Then suddenly, inexplicably, as she related to us, the plants all died. Not a potato had been produced, she sadly told her friends. Surveying the scene of desolation, she tripped on a bulge in the soil. And what did she dislodge? A potato big as a softball. Hmmm. She examined the soil more closely. *Why, the ground was full of potatoes!*

CHAPTER 2

Pastoral Economics

A farm is like a man—however great the income, if there is extravagance, but little is left.

Cato, *On Agriculture*, circa 200 BC

Borrowers are nearly always ill-spenders and it is with lent money that all evil is mainly done and all unjust war protracted.

John Ruskin, *The Crown of Wild Olives*, 1866

To make a go of contrary farming, it helps to forget everything you think you have learned about present day cost accounting. Not that you are going to ignore good business practices, but you should approach these from a different point of view: the view of pastoral rather than industrial economics. There are today no pastoral economists in our universities, governments, or businesses as far as I know. Harold Breimeyer, retired from the University of Missouri, comes close to being a pastoralist; in talking about how economic and political policies favor large-scale corporate farming, he once said to me in frustration, "Sometimes I wonder if our economic policies are not deliberately manipulated to get rid of family farms." He then went on to philosophize:

> Historically the trend to larger farmers does not reverse itself. Ancient Rome went from family farms to an estate system of agriculture with no return. Historically, small 'yeoman farming' starts when new nations emerge and is superseded as the value of the land grows. It may be logical or it may be paradoxical, but the fact is that yeoman farming lasts best where it is not very profitable in the strict capitalistic sense of the word. When farming yields a surplus above subsistence,

land is coveted. The more it is coveted, the more it flows into the hands of the people best able to buy it or powerful enough to take it. If current trends are not reversed by policy decisions, then at some date not far in the future, an alarm bell will sound about what will have already happened. It will be too late to do anything about it. We will be on a direct course to a concentrated structure of agriculture, whether through contractual integration, corporate conglomeration, or some form of semi-feudal barony. The family farmer will be extinct.

Scott Nearing, of homesteading fame but an economist by prior career, also understood the stand-off between pastoral economics and "price-profit" economics, as he called it. In *The Making of a Radical*, he wrote about his maple sugar business: "We were not trying to make money; that is a game in which the sky is the limit. Instead we asked ourselves: 'what is the least cash we can get by on during the next twelve months?' When we fixed that amount, taking into consideration all our plans and purposes, we produced enough of our cash crop to equal that amount and to provide a safe margin. Then we stopped production till the next budget year." He then went on to explain: "We aimed to be as free as possible from the market and from wagery. Price-profit economy presupposes the exchange of labor-power for cash; the payment of a part of the cash in taxes in exchange for regimentation, and the expenditure of the remainder in the market for food, clothing, gadgets and other commodities. The individual who accepts this formula is at the mercy of the labor market, the commodity market and the State."

The best twentieth-century articulation of pastoral economics that I have found was written in 1940 about England by an Englishman, Lord Northbourne, in a book called *Look to the Land*, now scarce and long out of print although it reads as if written next week. It is difficult for me to select a short passage from this book for quotation since the entire argument makes such a cohesive and inseparable whole, but I will try:

Urban and industrial theories and values have supplanted the truer ones of the countryside. These true ones survive mainly as a sentimental attachment to country life and gardening. Is the romance of country life really only a poetic survival of a bygone age, not very practical because there is no money in it, or is that romance something to which we must cling and on which we must build? Is farming merely a necessary drudgery, to be mechanized so as to employ a minimum of

17

people, to be standardized and run in ever bigger units, to be judged by cost accounting only? Or is the only alternative to national decay to make farming something real for every man and near to him in his life, and something in which personal care, and possibly even poetic fancy, counts for more than mechanical efficiency?

The answer, for Northbourne, is a resounding affirmative, as he goes on to answer his question, the question that I believe is central to the survival of our civilization.

> Mechanical efficiency is all very well—it is good, but life can be sacrificed to it. Mechanical efficiency is the ideal of materialism, but unless it is subservient to and disciplined by the spirit, it can take charge and destroy the spirit. In life, though not in mechanics, the things of the spirit are more real than material things. They include religion, poetry, and all the arts. They are the mainsprings of that culture which can make life worth while. Farming is concerned primarily with life, so if ever in farming the material aspect conflicts with the spiritual or cultural, the latter must prevail, or that which matters most in life will be lost . . . Farming must be on the side of religion, poetry, and the arts rather than on the side of business, if ever the two sides conflict.

Trying to think, much less act, outside the confines of industrial economics is very difficult because none of us realize how embedded in our psyches and our lifestyles are its tenets. It is like expecting a staunch, lifelong Christian, from a family that has been Christian for twenty centuries, to think objectively about other religions. If I were to say, for example, that capitalism and socialism are in practice more alike than they are different, most people, certainly most economists, would object strenuously because we have been taught that the two are absolutely opposed to each other. But both accept the same basic *money* system that the industrial revolution encouraged: 1. the use of pieces of paper or metal to represent real goods; 2. the acceptance of interest on these pieces of paper and metal as essential to "growth" (keep in mind that hardly four centuries ago, in a pastoral world, all interest on money was considered usurious and immoral); 3. the right of authority to manipulate interest rates—changing the definition of usury whenever self-serving authority desires it; and 4. the necessity of an expanding credit system that a government or bank can turn on or off at will in an effort to cover its own ass. Both capitalism and socialism, in other words, use money to centralize control over society. They differ only in who the central authority should be: socialism wants it to be the public sector and capitalism wants it to be the private sector.

Contrary farmers have no choice but to live and work within the capitalist/socialist economy that society now accepts with such naiveté. But if success is to be achieved on our homesteads, we must separate ourselves as far as possible from that economy. The first and foremost way to do that is to *borrow money only with extreme caution or not at all.* As much as possible, work with earned capital and non-cash assets like your muscles. This is especially true in agriculture and any other business activity directly connected to biological nature. There is something incompatible between biological systems and borrowed money. Farms are invariably lost because of over-borrowing; if they are not lost, they often are farmed in a bad way to get the quick money needed to pay interest. The problem is not so much the borrowed money itself, if only very moderate sums are borrowed, but the fact that rates of money growth (interest) seldom match rates of biological growth. An ear of corn never heard of money interest: it grows the same way whether interest rates are 6 percent or 15 percent. Nor is the weather going to smile benignly on you just because you borrowed a big wad of cash to buy a new tractor. You will have to pay that money back on a schedule that ignores the occurrence of drouths and floods, wind and hail.

Because nature and biology are not made the basis for the measurement of money growth, what happens is that for generation after generation, farmers borrow the stuff at a rate of interest that is not much lower, and is often higher, than their actual net return from farming. Like the sodbusters in old westerns, farmers still go to the bank in spring to borrow money and return in the fall to pay it back with maybe enough left over to buy their families a Christmas present or two.

You can borrow money presently at about 8 percent but it is difficult to get a net return on farming much higher than that. So farming has to be artificially subsidized to keep indebted farmers from dragging their banks and suppliers into bankruptcy with them. Since agriculture is fundamental to the entire economy (everyone must eat), government eventually finds itself subsidizing everyone. The only way to do that is with deficit spending. So now interest on the national debt ticks away like a time bomb, over half a billion dollars a day.

But how can a prospective cottage farmer not born to land or money ever own a farm without borrowing?

The answer is to start out SMALL. We think of a farm in the midwest as being at least several hundred acres in size and costing about $1500 an acre as of 1993. In other words this is something few of us cottage farm types can afford. I heard an economist tell a class of young people that if they wanted to be cash grain farmers today they had to have at least 1500 acres for a "viable unit." Worked right into that requirement for viability is a big chunk of income to pay off the colossal debts that such size is bound to generate. (Is it any wonder that bankers are so often found on the boards of trustees of agricultural colleges?) But such assumptions are meaningless from the standpoint of pastoral economics. A cottage farm of only an acre or two can, with high value crops, produce supplemental income which may be all the cottage farmer wants. (According to *New Farm* magazine, July/August 1990, Kona Kai Farms grossed $238,000 worth of salad crops on one half acre in Berkeley that year.) In traditional grain/livestock arrangements, one hundred acres devoted to astute husbandry can provide a decent living with pastoral economics, as the Amish prove over and over again.

The second precept in living pastorally in an industrial society is *do not try to make your entire livelihood from the farm, at least not at first.* Do like almost all our ancestors did, even in pioneer times: Pay for the land with a job not directly dependent on the farm's income. Even a casual reading of rural history in the nineteenth and early twentieth century shows that almost every farmer financed his initial land purchases by earning money in a hundred different ways—from teaching to blacksmithing to carpentry to working as a hired hand.

Today it may be even wiser to establish a way of life that permanently combines farming with another career or occupation. This has become true for almost all so-called mainstream farmers, a fact that agribusiness doesn't like to have publicized, so why should it not be so for small cottage farmers? Interestingly, Richard Body, a farmer and barrister in England who is also a member of Parliament, has reached the same conclusion for his country that some of us contrary farmers have long predicted for the United States: the present untenable situation will be superseded by thousands of new, smaller farms, most of them operated by people with dual careers. Two main factors point to this conclusion. First of all, these new farmers will be able to out-compete the very large commercial farmers for access to land, for reasons given below. Secondly, electronics allows such cottage farmers to pursue their other career in any rural place they wish to live. Writes Body, speaking of England:

> It's anyone's guess how many will make the move. Let us assume it will be no more than two million—that must be a very modest estimate. How many of them would wish to shrug off all their suburban values and immerse themselves in country living, must also be in the realm of surmise. While only a few might have the inclination, let alone the capital, to take over a 200-acre farm, it might not be unreasonable to expect one in ten to dabble in some form of agriculture. That is 200,000, which is much the same as the number who are farming today . . . Many will be content to produce only for themselves, but the 200,000 I have in mind will be engaged in serious part-time farming and like today's modern farmers, they will not be motivated by self-sufficiency but engaged in securing a supplementary income from the market (*A Future for the Land*, edited by Philip Conford. Green Books, 1992).

With another job, borrowing can be avoided or kept to a minimum. Here are some ways that my neighbors and I have found successful:

Rarely borrow money specifically for farming. Always have a dual use or intention for that money. If you have a job and have any savings sense, you can, for example, borrow for a country residence and pay it off out of job earnings. You obviously are going to have to pay for housing one way or another. The trick for the prospective cottage farmer is to pick a house that has land with it. In villages and rural areas there are reasonably priced homes that come with an acre or two, and that is plenty of land to start with. In many instances, there are humble rural homes on twenty to fifty acres that can be purchased for less than the price of an average suburban home on a quarter-acre lot. We bought our original twenty-two acres as bare land (no house or barn) for $14,000. Had we purchased just an acre lot for a house in the city where I was

21

previously working, it would have cost us at least $10,000. In a way I bought a farm for the price of a house lot that I would have had to buy in any event.

A young man I know, with a cottage farm in his vision of the future, bought a duplex in a village for $45,000. (The same house in a city would have cost him $100,000.) He and his wife live in one half, and rent out the other half. The rent covers most of his house payment. At the same time he is making improvements on the house using his own labor. He will have the place paid for in about a third of the time he would have spent paying off the mortgage on the $100,000 home in the suburbs, and he will do it without dipping into his monthly savings. In ten years he can rent out both halves of the duplex, or sell the house and leverage the money into a cottage farm.

I avoided borrowing for an additional ten acres with "cooperative buying." When a nearby ninety-acre farm came up for sale some years ago, four of us cottage farmers bought parts of it. Each of us could afford to buy ten- to thirty-acre parcels, but none of us could afford the whole thing.

Just recently two other cottage farmer neighbors bought sixty acres that lay between them—thirty acres each. That much each could afford without borrowing a whole lot of money. More significantly, they could outbid the large cash grain farmers who had to weigh the price purely against the per-acre costs of their commercial operations, which as we shall see are higher than cottage farm costs. The commercial cash-grain farmers could not justify any other consideration that made the land more valuable to the two cottage farmers: residential value, recreational value, husbandry value, or market gardening value.

Another form of cooperative buying is co-ownership. Three or four families might buy a hundred-acre farm together and all live on it without subdividing it. However, human relations problems can arise with this arrangement. Once upon a time there were two good buddies who bought four acres of woodland and pond as a vacation spot for themselves and their families. They had a falling out and wanted to divide up the property. But how do you divide a body of water between two owners? Yep, that's what they did. Put a rope right across the middle of the pond: one family on one side, one family on the other. Fortunately, with time, the feud mellowed.

Many cottage farmers start by buying a plot of bare land out of savings—two, three, maybe five, maybe forty acres. Then they slowly, in spare time, build a house and barn on the property, often, like Scott and Helen Nearing, using the rocks and lumber from their property for their buildings. They in effect "borrow" their own labor, time, and raw resources, rather than borrow money.

Some farmers have gotten out of financial trouble (the result of borrowing) by selling part of their farms. Often they find that by focusing their time on a smaller number of acres, they can farm better and make almost as much income, which they can keep rather than paying to the bank as interest. Ken West recently visited our farm and gave me permission to use his case as an example. Ken, in northern Montana, was grain-farming four-hundred acres and going broke. He reduced his size to a hundred acres, got out of debt, and began raising organic oats for oatmeal as well as continuing to keep his small herd of llamas. Yeah. Llamas. He says he sold one recently for over $20,000.

All five methods and others you may think of will only be successful if combined with a frugal lifestyle. Although few people seem willing to admit it, it is easier today than ever to save money. The more things go up in price, the more you save by not buying them. Most of us could be saving more but think we must spend it on "necessities" that are really luxuries.

- *Don't buy gadgets you don't need.* Why buy garage door openers unless you are disabled? Ice crushers are unnecessary unless your hands are crippled with arthritis. We have one, and it is just dumb: you can crush ice much more handily by putting a bunch of cubes in a cloth bag, and pounding them with the flat of an axe blade. Leaf blowers are rather stupid. I saw three workers on the lawn of a public institution last week, each armed with a blower, trying to corral a flock of about twenty-five leaves into a pile. What I wouldn't have given for a camcorder . . . but that's another doo-dad you don't need. The more kitchen appliances that people on the verge of bankruptcy own, the more they eat out. Don't buy clothes you don't need. A good suit can cost eight hundred to a thousand dollars today. A thousand dollars buys an acre of land that, in the right hands, might make an entire living. Let those who put their faith in fancy threads laugh at your jeans. Bury them in their thousand dollar suits.

- *Be especially astute about buying automobiles, your biggest cost next to housing.* If you borrow money to buy a new car, you will pay for it at least twice because of interest on your debt. Owning even the cheapest new car will cost you $2000 to $3000 a year out-of-pocket, if you use your own money. If you use the bank's, that car will cost you $4000 to $6000 per year. People who pay on car loans all their lives will spend a hundred thousand dollars in interest alone. Much of that money could otherwise be in a savings account making you money, not losing it. If you have only enough cash to buy a $1500 car, be content. Better yet, drive a $1500 pick-up truck, because that is the most useful piece of equipment a cottage farmer can own.

- And that brings up another excellent way to save money. *Learn how to fix cars and tractors yourself.* The labor charge at the garage for car repair is something like $30 an hour. Being able to keep older piston-ringed roarers in repair enables you to avoid one of the greatest costs of living: depreciation on newer piston-ringed roarers.

- *Where you live can be the source of your greatest savings.* If you decide to live in the city (tah dah) where the salaries are higher than in rural areas, don't bet that you will come out ahead. A newer nice house, but nothing terrific, presently costs $150,000 in Columbus, Ohio, to consider a place I am familiar with. Some sixty miles away, in our village, a house similar to it, perhaps a little older, can be purchased for $90,000. A humbler village house, older yet, but still well built though possibly requiring some modernization or renovation, can be purchased for $40,000. Think of the enormous savings in interest over a lifetime, by going the cheaper route. The best kept secret in America is the bargain that awaits the wise home buyer in smaller towns and villages, the most pleasant places to live in the United States.

- *Receiving interest on money is a far, far different story than paying interest out, as the rich know so well.* When the Pontiffs of the late Middle Ages realized how lucrative exponential money interest could be, they quickly changed the Church rules prohibiting usury. To get an idea of how fast your savings can mount up, remember the rule of 72. You can roughly calculate how soon savings will double by dividing the interest rate into 72. If the interest rate is 8 percent, $5000 will grow to about $10,000 in nine years—72 divided by 8. Even more roughly, you can generally figure over a period of twenty years, during which interest rates will gyrate up and down at the whim of the Federal Reserve, that savings double about every ten years.

Since you pay a higher rate of interest on borrowed money than you get on saved money, you can understand, by the same rule of 72, how borrowers most often stay poor. You can also see why a person of humble circumstances who saves even $1000 a year, at an average 6% for forty years, has a very nice nest egg for retiring to a cottage farm in his or her mature years. A friend of mine, doing exactly what the expert farm magazines advised against, farmed all his life on rented land, never making much money but spending even less. He bought a 180-acre farm when he was sixty three for cash with plenty left over along with farm income and Social Security to live comfortably for the rest of his life. Essentially his "secret" was simply to stay away from borrowed money and save a little every year. As I recall, he borrowed only once: to buy his first (used) farm machinery. His other reward for always avoiding the luxury that borrowed money promises has been, as he puts it, "not ever having to kiss some sonuvabitchin' boss's ass."

- If your goal is cottage farm contentment, *some of the most unnecessary money you can spend is for a college degree.* The colleges would have you believe that a college degree means more money in your lifetime, which is precisely the wrong reason for trying to get educated in the first place. In the second place, it ain't necessarily so. Your other career may demand college certification, but if not, it really is a little foolish to spend a lot of money for information that is now easily available from other sources. Just read every book and magazine you can lay your hands on or can access through your computer or through inter-library loan. A College of Agriculture degree (in agricultural economics, for example) is especially unnecessary for your future as a cottage farmer (or any other kind of farmer). There are easier and cheaper ways to gain that knowledge. The $40,000 to $60,000 you spend for a college degree, which goes mostly for the inflated salaries of professors who are more interested in their research than in teaching, will do you more good if invested in land or a business, or good tools, or a savings account or books and modem hookups to libraries or to computer bulletin boards. Add to that the money you could be earning instead of sitting in sterile classrooms, out of which you could save another $5000 a year, or $20,000 for four years. That's a hundred thousand that could be doubling every ten years if saved, or possibly making a living for you if invested in a cottage business.

In the early years of saving, the eventual reward seems hopelessly far off. But if you stick to the habit of setting aside some money each month or invest that money in land or tools that produce something for the long term, you will be rewarded not only in a bit of financial security but in contentment, because the way of life you must follow precludes endless striving for Things. Once you realize how much more precious independence is than all those Things, saving is easy. If your spouse doesn't agree, you have the wrong one for cottage farming.

Financial Yardsticks in Cottage Farming

After learning how to "think small," the next rule of pastoral economics is to measure the value of products in *human* terms, not in financial terms.

"Why do you raise sheep when there's no money in it?" I asked a fellow shepherd.

"Well, there's a little money in it," she replied. "But the real reason is that my sheep make me happy."

Even suggesting that happiness should have a monetary value draws gales of laughter from those who think of themselves as hard-nosed economists. But as a matter of fact, happiness does have a very practical value because stress and unhappiness are known to cause health problems that can result in astronomical medical bills. Industrial economists have no way to measure that value so they ignore it.

I have admired for years the home economy and contentment of a couple who live in a tiny village near my home. I asked them once how it was that they remained in such good health and active life far into their retirement. The husband, working in his woodworking shop and listening to a Mozart tape, smiled. "We have always made it a point to avoid stress as much as possible," he said.

In industrial economics, sheep have two measures of value: as mutton and as wool. What are the additional measures of value in pastoral economics?

1. Money earned from one's labor can be counted as a profit, not a cost as it is in industrial accounting. Shepherds consider their time spent in raising sheep and any money derived from that time as their wages—their profits so to speak. On the other hand, for a large-scale farmer hiring people to do his work so he can expand in an industrial economy, labor is an expense, a necessary nuisance that the Lord of the Manor must pay out before he can count his profits. Of course, when I hire work done, as for sheep shearing, I assume, to that extent, the financial situation of the industrial economy. It costs me more in out-of-pocket expense to hire than to shear the sheep myself.

   (There is an interesting effect of industrial accounting in this regard that I can't resist pointing out here. Because labor is considered a cost in industrial accounting, and because farm labor has not been fully assimilated into industrial economics—has generally not been unionized or organized in any other way to bargain for a fair wage—farm work is assigned a low value: $7 an hour is the figure agricultural economists currently use. As a matter of fact, I know laborers who drive one hundred thousand dollar harvesters for large scale farmers today, or who milk the cows in thousand-cow dairies today, who make less than that. Economists who place the value of farm labor at $7.00 per hour while assembly line auto workers make $20.00 an hour, reveal how deep-seated is the industrial prejudice against farming. And with reason: if farm workers were paid on the same level as urban workers, industrial farming would long ago have collapsed or food prices would have doubled.)

2. When the shepherd eats his own lamb chops, the value of the meat quadruples over what the stockyards pay because that part of the cottage farm production that is used by the cottage farm household can be valued according to its retail prices. The time I spend producing food for my own family is more profitable per hour, with this kind of accounting, than the time I spend writing. And that does not take into account the probability that our home-grown food is more nutritious.

3. Wool sold directly to a local spinner rather than to the conventional wool market can bring twice to three times the price. If a shepherd processes the cottage wool into clothing, purses, or rugs, and counts the labor as income—that is, as a foregone or avoided expense—the value of the wool goes up even more.

4. Even when I barely break even with my lamb and wool on the conventional market, I "profit" because the sheep mow and fertilize my fields for me as part of my non-herbicide weed control program, and keep fencerows relatively clean of weeds.

5. Sheep manure is second only to poultry manure in fertilizer value, and the mere accounting of the amount of nitrogen, phosphorus, and potash in that manure does not begin to add up to its total value for soil fertility.

6. The value of keeping sheep, as opposed to letting some big-tractor farmer turn all the sod land required by sheep into erosive surplus corn, is immeasurable. The sod land will not erode; the sheep will provide the fertilizer and replace the cost of mechanically harvesting those acres, and thus improve the sustainability of the land while decreasing the chances for pollution.

7. The value of a husbandry-driven infrastructure of small cottage farms across the whole nation is also incalculable. If healthy, such a rural culture could mean who knows how many people retreating gladly and willingly to the countryside, relieving the population pressures that are turning cities into heat sinks of human frustration.

Spreading out the population to share the life of shepherd and cowboy would hopefully generate a renewed emphasis on traditional rural virtues and give families a reason to work and play and love together again.

I sometimes entertain myself with a crazy vision of Ohio. From Lake Erie to the Ohio River, the land is quilted into meadows, tree groves, and playing fields, the latter kept manicured mostly by the grazing animals upon which the local economy would be based. In the groves, pastoral "factories" would be built—home/workshop combinations where most of the society's basic manufacturing would take place. Instead of staring at flow charts and spreadsheets all week, today's office workers could sit in their home workshops and spin and weave world-famous Ohio Persian rugs or construct world famous solid walnut furniture, making a living by producing something they could be proud of. Instead of gathering on street corners to await in boredom the day when they can join the real world of meaningful work, teenagers could have just as much fun gathering in field corners where they would also be keeping an eye on the flocks. Every weekend the towns would host food and craft fairs along with sports tournaments. Golf addicts could play 7000 holes from Cleveland to Cincinnati and never leave the sheep pasture golf courses. Champion basketball and football players could become champion sheep shearers and hay bale slingers in the off-season and thereby contribute something useful to society while staying in shape. Teams of former factory workers could travel the grove and grass landscape, turning excess lumber into houses and home fuel. They could build small lakes and ponds for water storage, fishing, and swimming. We would all live eventually in a region-encompassing natural preserve. Good T-bone steaks and lamb chops and fresh fish would become as cheap as White Castle hamburgers. Pollution would diminish drastically because long-distance truck transportation would lose much of its economic base and long-distance vacation driving would lose much of its allure.

Well, I can dream, can't I?

There is another aspect to such romantic visions of a pastoral economy that is not at all as impractical as the foregoing sounds. This aspect speaks to a situation too far in the future to articulate cleanly yet. But as national columnist Richard Reeves has been thoughtfully suggesting, we may be looking at a future where there is not enough full time, *salaried* work to go around. It is scary enough when big companies lay off workers right and left, and others flee to third world countries in search of cheap labor. But when Procter and Gamble says it must lay off workers not because the company is losing money but so that it can continue to make money, we may be hearing the first announcement of the end of the industrial society—the end of a society based on an ever-increasing scale of wages and an ever-enlarging rate of production. Whether one looks at that possibility in the practical short term—What will I do if I can't find a steady job?—or in the theoretical long term—What if the number of good paying jobs, especially white collar jobs, continues to decrease?—the cottage farm and workshop begin to look more and more appealing as a safe refuge. Within hardly a two-mile radius of our farm, in a county with a population of only 26,000 people, in a landscape of widely dispersed houses and farms, there are seventeen home businesses, not counting farming. Kathryn Stafford, associate professor of family resource management at Ohio State University, who participated in a recent nine-state survey of home businesses, informs me that such numbers are not unusual today. Most of the businesses in our neighborhood generate supportive income, not main income, but most of them are also hedges against possible interruption of the family's main source of paid wages. I don't describe those of us maintaining our livelihoods at home as having a sentimental yearning for the past, but rather a very practical vision of a post-modern pastoral economy where real goods count for a lot more than money.

If Money Talks, Make It Speak Your Own Language

The industrial economy has been methodically closing the doors of opportunity to pastoral farming for a century. The main latch and lock, so to speak, is the policy of giving price breaks and other preferential treatment according to quantity in purchasing supplies and selling products. The more you buy, the cheaper per unit industry can price it to you. The more you sell, the more market opportunity becomes available to you. But this policy only works if there are many small purchasers and producers paying the higher price so that industry can offer price breaks to the big guys. If all the buyers and sellers become large scale, there can be no price breaks for any of them. Thus, the small business subsidizes the big business.

For example, if you buy small amounts of fertilizer or seed or chemicals, you will pay higher unit prices than the large-scale buyer. But as all cash grain farmers expand in size, the difference decreases steadily, while the value of manure and saving one's own seed increases.

Since smaller farm machinery dealers are going out of business right along with smaller farms, the survivors are using a new variation of the quantity-price-break tactic. They are offering break-even prices to the big cash grain farmer if he trades in several big-ticket machines every two years or so. The dealer then tries to make a profit on reselling the trade-ins at padded prices to debt-burdened farmers who desperately need to update their equipment but can't afford the cost of new equipment.

This ploy is not working. Farmers looking for used equipment are beginning to shun the dealers and dealing directly with each other. Last winter I was fortunate enough to be able to buy a good used 50-horsepower tractor, with a loader on it, for about $3000 less than similar models at dealerships. A friend of mine bought a used 150-horsepower tractor at private treaty for something like $15,000 less than what he says were his alternatives through a dealer. Then he turned around and sold his smaller tractor to another farmer for about what the dealer would have given him on trade-in.

Quantity has been given the edge when farmers sell their production, too. Milk haulers are not interested in driving another ten or twenty miles to service a five- to ten-cow dairy herd anymore. Their hauling profit is based on volume per mile. Yet only in a few states can the small dairyman sell unpasteurized milk directly to customers even though good dairies have proven that with modern methods and equipment, milk does not need to be pasteurized anymore. The hollowness of the prevailing regulations are readily apparent to anyone who buys milk at Young's Dairy near Yellow Springs, Ohio. Young's has been selling unpasteurized milk for forty years with no problems. Their customers actually prefer it. Young's can do it because the dairy was selling unpasteurized milk before the law went into effect and so can continue because of the grandfather clause. But until "health" regulations are changed, newer, smaller dairies don't have this avenue open to them and therefore can't pass on to consumers the lower-priced milk that they are capable of producing for direct retail sales. The name of the game against them is monopoly, not consumer health.

Until the tenets of pastoral economics are again recognized as integral to a sustainable society and are reintroduced into mainstream economics, (or even if they never are), cottage farmers must learn how to live with the disadvantages that small size brings and concentrate on quality, efficiency, and cooperation to make up for it. They have to generate their own underground economy, which is what they are doing when they trade used tractors among themselves. Some farmers who have learned that they can get along with each other well are sharing the ownership of an expensive machine rather than each owning one.

Along with sharing equipment, cottage farmers must learn to trade work instead of money. Our grandfathers and fathers did that as a matter of course. Just recently I traded three days of work helping neighbors Pat and Steve Gamby fill their silo for some used fencing that would have cost me a couple hundred dollars if I had to buy it new. The "profit" in the deal was a remark that Pat made and Steve agreed to, which I treasure beyond all wages: "The only way small dairymen like us are going to survive is if the price of milk goes *down*," she said. "Then it would be the big boys going out of business."

The Amish have been geniuses at living pastorally within an industrial economy. They produce at low, horse-power costs and sell at the high, tractor-power prices all the time. Where they might spend a dollar per bushel out-of-pocket to produce corn, the "English" farmer, as they call us, might spend $2. This is the main, though unconscious, reason why some of the "English" are prejudiced against the Amish. The Amish aren't playing the game "fair."

Actually when any financially conservative contrary farmers put off spending in order to save money rather than borrow it, they are not playing the game "fairly" either, in the eyes of most industrial economists. They are saving in the pastoral economy but acquiring interest in the industrial economy. If *someone* didn't borrow all that money (and buy all that junk), savings would not "grow" by exponential interest rates, the industrial economists point out. It is our duty to spend and consume like crazy to keep the economy booming.

But if there were no interest on money *anywhere*, savings would not have to "grow" to "keep up" with "inflation" and the economy would not have to "boom" to recover from a "bust." A gentle, continuous buzz would be sufficient.

Among themselves, the Amish cultivate an effective underground economy that the rest of us could learn from if only we would. At a barn-raising I attended, a huge timbered barn was erected to replace one that had been blown down by a tornado. This whole process cost the Amish farmer about $30,000, most of it "borrowed" interest-free from the community's own insurance fund, into which all church members pay. Many Amish do not charge interest on loans to each other, and when they do, the rates are invariably cheaper than bank rates. Moreover, had the barn been built by contractors and workers in the mainstream economy, I was told it would have cost $100,000, if workers could be found at all to do the kind of skilled mortise and tenon construction that the Amish employ.

Also, most Amish sects do not build churches but use their homes for church meetings. Nor is a hierarchy of ministers possible because bishops are chosen by lot, and they continue to work their farms or businesses like all the rest. How many zillions of dollars are saved by not being burdened, as most other Christian, Muslim, and Buddhist sects are, with a huge superstructure of drones and stones.

Nor do the Amish finance retirement homes. (They don't accept Social Security.) Grandparents in their old age live in houses on the farms that their sons and daughters take over from them. This of course means much sacrifice sometimes in caring for the old folks in their final days, but how much money is thus saved is probably beyond accounting. More importantly, the panic with which we English now face old age, as costs catapult beyond what we will ever be able to afford, is perhaps the best proof of all that an industrial economy doesn't work.

All small farmers worry about where affordable smaller machinery will come from in the future. The Amish have learned how to repair and restore old equipment almost indefinitely. Alton Nisely, for example, operates a repair shop in Holmes County, Ohio, that uses very sophisticated, high-tech machinery to re-tool—and even manufacture from scratch—parts for the old horsedrawn machinery that the Amish favor. Wayne Wengerd, not far away, operates a small factory, the Pioneer Equipment Company, where he manufacturers new horsedrawn plows and other equipment that mainline manufacturers no longer make. Interestingly, he tells me he has just as many "English" customers nationwide as Amish.

Finally, in building a new trade complex between cottage farmers and their suppliers, many other cottage industries need to be generated and supported. These small industries are often called backyard businesses. For example, we are fortunate enough in our neighborhood to have what I call a "freelance mechanic" who is a genius at his work. He used to be the national troubleshooter for a large farm machinery company until, as he puts it, he "got tired of going to meetings all the time." He has found a profitable specialty in the cottage farm economy: he puts rebuilt alternators on old tractors to replace original generators, so that the tractors start almost as easy on a cold morning as a horse. He has more work than he can handle.

Country welding shops are another good example. One of my nephews down the road can turn a pile of scrap iron into just about any tool a contrary farmer might dream up. He is looking forward to the day when welding and repair will be his little farm's main source of income, just as writing is our farm's main source.

A caveat. In the absence of an infrastructure of trade for cottage farming, the temptation is to try to do it all yourself. Make the equipment. Operate the farm. Become your own farm produce retailer. Many articles and books will tell you how to do everything yourself and "make more money." Don't believe it. Most of us are already working a second job and two careers are enough. Or, if you have graduated to full time farming with husbandry (that is, raising animals), you will find that your time is fully taken up. Oh, you will surely cobble some of your own equipment together: that's part of farming. Or you may operate a small, temporary roadside stand, or take produce to a farm market occasionally. But if you try to complete the trade complex circle by operating a retail business (for instance, a butcher shop or full time fresh food store) in addition to farming, you will create havoc in your life. Not only is there not enough time to do both well, but more often than not a farmer hasn't the temperament to be a storekeeper, and vice versa. Most people drawn to farming do not like selling and so are not good at it. Much better to connect with someone who understands and likes selling and let him or her make some money too. And when you need a special tool, pay a mechanic to make it for you. You will be happier, will make more money concentrating on your first love, and will help build a community of independent but interdependent people.

Another way to start generating such a community without overwhelming yourself with work is to look into the possibility of "community supported agriculture" where you sell your products through subscription farming, pick-your-own, or other cooperative ventures that intimately involve the participation of your customer-clients (see chapter 3).

A Parable from the Industrial Economy

In the beginning, the Lord God of the Economy saith: Let there be the General Store and it appeared on every corner and crossroads in America. And the Lord God of the Economy saw that it was good. The stores served almost everybody within walking distance of their homes. Even far out in the country the people were never more than a mile from an ice cream cone or a new pair of bib overalls, about all that they required of industrialism.

But these scattered, decentralized, Mom and Pop stores were not industrially "efficient" and the Lord God of the Economy became displeased with them. "Because thou has not hearkened to my commands, I will cause thee to raise up a son who shall be the death of you," He said to Mom and Pop. And He put forth His hand and lo, a bright young entrepreneur, fresh out of college, built a new, self-serve store on the edge of town to catch the rural trade coming in and the urban trade going out. By buying in slightly larger quantities and making the customers serve themselves, thus saving on labor, he sold slightly cheaper, or so it seemed, and ran Mom and Pop out of business.

And the Lord of the Economy looked upon what His servant had done and said: "This is good. Mom and Pop were tired of storekeeping anyway. Let them playeth golf for the last twenty years of their lives and pass happily into paradise."

But in time, the servant became complacent about profits and in anger, the Lord God of the Economy raised up a chain store which, by the same quantity pricing, ran Mom and Pop's son out of business. And the Lord God of the Economy saith: "This too is good. The consumer hath gotten a better price, so now she can buyeth more." And the evening and the morning were the third day.

But farther out on the edge of the village, the waters of competition did gather together and upon the dry land of greed, behold, a shopping center did appear which harkened yet better to the precepts of the Lord God of the Economy and so drove the chain store out of business too.

And the Lord of the Economy saith: "This is even better. Look how the economy heateth up and provideth more jobs."

But lo, the people still did cry out for more Things to buy. So the Lord God of the Economy stretched forth His hand and behold, the firmament parted and a Mall appeared out beyond the traffic mess created by the shopping center. Now there was so much selection for so few drachmas each that the Things barely paused a year in the consumers' households on their way to the landfill.

The people only groaned louder in their travail at constantly having to replace old junk with new junk to keep the economy going. So the Lord God of the Economy cast His cape over the land, and there appeared a Super Mall in the city sixty miles away. How grand it was with trees and shrubs and waterfalls and not a drop of rain, and with the sound of the zither and the cymbal to spend by. So the people did drive there to perform their rites of shopping and the hometown mall became as deserted as Main Street.

But there was still much weeping and gnashing of teeth because earnings did not keep pace with the prices. So the Lord God of the Economy spread his hands over the land yet once more and there appeared a Super Outlet Mall far out in the middle of nowhere but within two hours drive of everywhere. The people bowed before the God of the Economy and abandoned not only their hometown stores but those in the larger cities now, to motor on to the Outlets that they might pay $10 for the privilege of parking, and walk three miles through the citadels of materialism to save $15 on a pair of shoes they would not otherwise have needed. Such a bargain was possible because they had spent an entire day and $15 in car expense to get within three miles of the store. And the evening and the morning were the sixth day.

And so the generations passed even unto this day. Eventually the people ran out of money needed for fuel to drive two hours to the Outlets to save money. A consumer paused, as she stumbled tiredly through the maze of shops, most of them boarded up. "This Outlet remindeth me of something," she said. "It remindeth me of Main Street. Why don't we all just go home and open up Mom and Pop stores that we can walk to and give Mom and Pop something to do besides play golf and bitch about their Social Security checks?"

And so it came to pass. A whole new generation of songwriters gathered on the cracked pavements of the deserted parking lots and sang sentimental post-Rap ballads about the good old days of the malls when everyone had plenty of borrowed money with which to buy everything except happiness.

CHAPTER 3

The Garden is the Proving Ground for the Farm

Today's organic gardens are the experimental plots for tomorrow's agriculture. In fact, the typical farm of the future may be simply a very large garden.

Robert Rodale, in conversation with the author, 1984

In 1965 my wife and I rushed into serious gardening on backyard land where commercial farmers would have feared to tread. Much of the topsoil had washed away from this ridge in an earlier farming era—we could see the depressions of the old dead furrows in our grassed acre of back lawn. Moreover the soil was underlain with almost solid rock less than two feet below the surface. The garden dried up when rain was scarce and even with plenty of moisture, plants did not grow luxuriantly. We turned to mulch gardening, a practice just then receiving serious attention. Using our own tree leaves and grass clippings plus a truckload more that the village gave us annually, we began mulching everything we grew, from rows of corn to beds of flowers. The results of mimicking nature's way of growing plants were quite dramatic after the leaves and grass clippings began to rot at the bottom of the layer of mulch. Not only did the leaves keep the soil from drying out during periods of drouthy weather, but they controlled weeds, prevented erosion, provided plant nutrients, and increased the organic matter content of the soil. I even finally used leaves in place of cultivation: instead of spading up a new section of sod, I covered it with leaves the first year, which killed the grass, and then transplanted tomatoes and other potted plants down through what remained of the rotting mulch the second year, adding more leaves as the growing season progressed. By the third year the soil was easy to rotary-till for seeded crops, or I could simply plant seeds in the duff of leaf compost that had formed on the soil surface and continue to add leaves when the garden plants got a few inches tall.

I was so amazed at how well mulch gardening worked without fertilizers or herbicides or irrigation or even hoeing that I became an ardent supporter of organic methods, and wondered aloud at the offices of *Farm Journal*, where I was working, whether farming too might not be done "organically." No, said my peers, all ardent supporters of chemical farming. "Organic methods are fine in city gardens, but aren't practical on commercial farms." I wonder how many times I heard that.

Now fast forward the video cassette of my life to 1992. I am dining with a professor from the agricultural college at Purdue. Guess what. He is telling me about his successful experiments using city tree leaves as a soil amendment in farm fields now that yard waste is being banned from landfills. Obviously organic gardening methods are applicable to farms after all.

Purdue scientists were spreading leaves and grass clippings with a manure spreader ahead of conventional cultivation and planting, certainly better than burying yard waste in landfills and a lot cheaper than composting the stuff in large, centralized composting facilities. But applying yard waste to soil this way did not gain all the efficiencies of mulch gardening and ran the risk of temporary nitrogen deficiency in the crop if the leaves were incorporated into the soil too heavily—a risk not incurred when leaves are mulched on top of the soil and allowed to break down slowly. "I have long wondered," I said, hesitantly, for I thought I was suggesting a radical idea, "if a self-unloading silage wagon might be modified to apply leaf mulch directly to row crops after the plants were about six inches tall. You could straddle the rows with tractor and wagon and dribble the leaves out through the side unloading chute as you drove along, putting a mulch of desired depth down between the rows. That way you would gain the weed, erosion, and moisture control benefits of mulch gardening as well as the nutritional value of the rotting leaves." He nodded. "Yeah, we've thought of that too especially for sweet corn. Ought to be able to mulch four to five acres a day that way easily with present equipment, quite practical for small farms near cities. Another advantage, for sweet corn producers, is that the mulch would do away with mud at harvest time."

The moral of the story is that the proving ground for real change in farming has almost always been the garden. Commercial farmers are good at improving their existing technologies, but rarely do they initiate pivotal new practices because they are financially strapped to the mass market and can't afford to risk the possible profit loss of changing horses in mid-field. New agricultural ideas come from gardens where financial profit is not a necessary goal; generally these gardens are city gardens. Fresh new ideas in any institutionalized activity (and nothing is more institutionalized than agriculture except religion and education) almost always come from the outside. Writes Jane Jacobs in her provocative 1969 book, *The Economy of Cities*: "Modern productive agriculture has been reinvented by grace of hundreds of innovations that were exported from the cities to the countryside, transplanted to the countryside, or imitated in the countryside."

For example, alfalfa was a medicinal plant in Paris a century before it became a farm crop throughout Europe. Edward Faulkner wrote his revolutionary best-seller, *Plowman's Folly*, based on experimentation he did in a garden near Elyria, Ohio, not on a farm. It was city gardeners, not farmers, who, with ample supplies of manure from livery stables and street sweepings, brought real sophistication and efficiency to the use of animal manures for food production, as is amply clear from books like Benjamin Albaugh's *The Gardenette or City Back Yard Gardening*, published in 1915. It was urban influences, following the work of chemist Justus von Leibig in nineteenth-century Germany, that introduced to resisting farmers an agronomy based on chemicals. Today it is city gardeners, following scientists like Sir Albert Howard and Dr. Selman Waksman, who have introduced, again to resisting farmers, the notion of an agriculture based intentionally on biology. Leibig disproved the prevalent nineteenth-century notion that plants got all their food from humus. But in proving that plants "eat" minerals, not humus, Leibig went to the opposite extreme and demeaned the practical necessity of humus, and humus-derived nutrients, for a sustainable and efficient agriculture.

The whole organic farming movement, which now extends even to cotton, a crop once thought impossible to grow without toxic chemicals, was of course inspired by city gardeners. But organic gardens, which in my rural area are still sometimes viewed as the invention of "commie liberals," are not the only generator of change in this regard. Urban recycling is another. Waste paper has become an animal bedding of choice on many farms, not because farmers demand it, but because cities provide it at a cost that is cheaper for some farmers than straw. Composting, once strictly a garden practice, is now becoming a farm business too. Bob Birkenfeld and two of his brothers, who farm 3500 acres and raise 2500 stocker cattle a year near Tulia, Texas, have started a new business they call KBG Composters, making and selling compost with the manure from their own and neighboring feedlots. Birkenfeld is one of my favorite contrary farmers. He says there's not much profit in large scale farming and that he believes if he concentrated all his efforts on husbandry on his home 180 acres, he could do as well financially without the stress of the big operation.

With modern pre-treatment methods, urban waste managers have found ways to deliver an entirely safe, composted sewage sludge to farmers at a cost that is cheaper than conventional fertilizers. After initial hesitancy, farmers are lining up to get this material, which is also being used to restore strip-mined lands to forest and meadow. Unfortunately, organic farming organizations, after much debate, have disapproved the use of pre-treated, composted sludge on certified organic farms. To me this was a stupid move which I think springs from our silly fear of our own excrement. For ten years I have followed this debate and as a writer have worked closely with the leading sludge scientists. True, in earlier times, PCB-contaminated sludge was a remote possibility (though even then, ton upon ton of sludge was applied to Ohio farmlands with no problems), but with modern pretreatment and constant monitoring, sludge is, as USDA scientist Dr. Rufus Chaney says, as safe as any soil amendment or fertilizer can be. He once pointed out to me that there is more cadmium applied to farmland by way of commercial fertilizer in one year than in all the sludge ever spread. Farmer Gary Wegner, who has found sludge to be an excellent fertilizer and humus builder and erosion fighter for his dryland wheat farm in Washington, has been hired by the city of Spokane to promote the use of sludge in farming. He says he makes his most effective argument with a bottle of vitamin/mineral tablets. He reads from the label the vitamins and minerals that are contained in the tablets "from A to Zinc." "The heavy metals like zinc that people have been erroneously taught to fear in sludge are the very metals that they consume in health pills and which my soil is deficient in," he tells his audiences. Of course the Chinese would smile at this debate. They have been using human excrement on their garden-farms for forty centuries. The only problem it seems to have caused them is over-population.

Alan Chadwick, the famous advocate of French intensive and biodynamic horticulture, understood the importance of the urban garden in generating agricultural change, as is clear from a story that organic gardening's guru, Bob Rodale, was fond of telling. It seems that in the early 1980s, Bob and Wendell Berry (the two of them probably exerting more influence on agricultural thinking in modern times than anyone) went to California to interview Chadwick at the latter's famous garden in Santa Cruz. Chadwick, who, like Rodale, has since died, was a very opinionated and passionate gardener. His way of cultivating, beginning with double-digging and finishing with the hand rake and hoe if with anything at all, never the rotary tiller, was the only way one dared to garden under his tutelage, and he could be acerbic and even insulting in defense of his beliefs. (It is a good thing we never met because I think double-digging is a waste of time.) The university where Chadwick held forth on his gardening methods had by that time begun the logical progression: students of Chadwick had started a farm based on the master's garden methods. The farm annoyed Chadwick, perhaps rightfully, because it unwittingly drew attention away from the true source of its success—his garden. Without Chadwick's knowledge, Rodale and Berry were taken on a tour of the farm first, and had enjoyed a lunch with food provided by the farm. They then went on up the hill to Chadwick's digs, where the master was waiting for them. Before giving them a tour of his own, Chadwick invited the two famous progenitors of new farming ideas to eat a garden lunch outside the little hut where he lived surrounded by his plants. "When we told him we had already visited the farm and eaten there," recalled Bob, "he glared at us indignantly, and without a word turned around and stalked into his cabin. That was the end of the interview." Bob, who had an exquisite sense of humor, grinned at the recollection. "He never did come back out. We walked around a little while, embarrassed, and left."

Mechanical as well as agronomic technologies can move from garden to farm, strange as that may seem in an age of 300-horsepower tractors. The hand-pushed garden seeder is a good example. This simple, plastic two-wheeled seeder costs $50 to $70 depending on where you buy it, and will last a lifetime unless you run over it with your truck. Commercial market gardeners are now joining two or three units together for multi-row planting, doubling or tripling the amount of planting they can do in a day without fossil fuel. Two or three together are in fact easier to push than one. I thought it was my idea until I read Andrew Lee's excellent book *Backyard Market Gardening* (Good Earth Publications, 1993). Lee shows how to fit three Earthway planters together thereby increasing the amount of land that can be planted by one person in a day to at least two acres, which is about all many commercial vegetable farms need to plant at one time. Two people, each equipped with my two-row version, could plant three acres a day between them without undue fatigue and in a week's time plant twenty acres of corn, all that is necessary for the kind of small commercially viable controlled grazing farm this book advocates. Of course, the large operators who farm around me get a big kick out of watching me push my two-row planter planting corn, while they pull thirty-row planters across their fields. But since without their subsidies, they aren't making any money either, they don't laugh too loud.

Joining the seeders is extremely simple. I put my two together with three scrap one-by-two inch boards, each about three feet long. One board bolts to holes I drilled in the front wheel kick stands (with two seeders joined, they stand upright without need of the kick stands, so immobilizing the latter doesn't matter). I sawed slots in a second board which fits over the frames of the two seeders right behind the planting boxes. The third board bolts to the holes already drilled at the top of the handles, holes intended by the manufacturer for a fertilizer box. You could attach a board at this juncture even with the fertilizer box installed. I had thought that I would have to cross-brace the frame to keep it from cracking, but the three boards hold the two seeders very rigid. I can vary the distance between the two units (or three, or even four) by sets of holes and slots in the boards. I can quickly dismantle the frame if I want to use one seeder alone in the garden.

The rotary tiller is another machine that (except among Chadwick's followers) has gone from the garden to the farm. The first well-known example was the Howard Rotovator, first made in the 1950s, a large, heavy rotary tiller for farm tractors. It was touted as sustainable farming's replacement for the plow but only a few farmers have accepted it. The Lilly Roterra is a more modern adaptation of the rotary tiller idea, but it too has not won over many farmers and for the same reason: primary and secondary tillage with a Rotovator (or any other rotary tiller) was considered too slow compared to other methods. Also, and here I agree with Chadwick, most rear-end tillers spin so fast that they can harm soil structure if overused. They can also create a hard layer of soil at normal tilling depth just like a plow or disk can, if deep rooted crops to perforate the hardpan are not kept in the crop rotation.

A more promising adaptation of the rotary tiller to farming is as a weed cultivator, at which this machine has no equal in the garden or field. Shovel cultivators can slide around weed roots, but no weed can escape the tiller's chopping action. Coming on the market now are multi-rowed, hydraulically and PTO-operated tiller cultivators mounted on farm tractors. They look like big steel spiders whose legs each end in a little tiller. This is one answer to the commercial farmer's new interest in mechanical weed control now that weeds increasingly exhibit resistance to herbicides.

Gardens act as incubators for the farms of the future by exploring new marketing methods as well as field production strategies. The new generation of commercial market gardeners nearly all began as backyard gardeners enormously affected by the consumer desire for low cholesterol, pesticide-free, fresh food. Ward Sinclair, Pulitzer Prize-winning writer turned market gardener, began selling his organic produce to fellow workers in the offices of the *Washington Post* where he worked. The number of customers grew. He tells me that he found selling food more satisfying than selling words, so he quit his job and started what is today one of the most successful commercial market farms in the East.

These new and often very contrary farmers were all drawn into direct retail sales in an effort to supply the city demand for organically clean food and to make a profit doing it, thus reconnecting the farmer with the consumer directly. This "reconnecting," so well exemplified by farmers' markets in the cities, makes for a healthy economic system. Human beings are mostly rather wonderful creatures in their *personal* relationships with their spouses, their children, their parents, their friends, their comrades at work. So when food shopping becomes a direct experience between the farmer and the consumer, mutual solicitude is invariably generated. The farmer wants the customer to be pleased, and the customer, so served, wants the farmer to continue in business. Farmers' markets thus represent one of the few remaining examples of genuine capitalism left in this country.

Compare that to the lowly-paid hired hand feeding ten thousand steers knee deep in shit at a big feedlot owned by millionaires for tax-dodging purposes. Do you think this worker gives a damn whether the meat he is helping to produce pleases some faraway urban consumer? Likely as not, he despises urbanites in general because he knows they make a lot more money that he does and can afford to buy the steaks his labor produces while he eats hamburger. This kind of farming is not free-enterprise capitalism but captive-enterprise socialism.

Pick-your-own selling is the granddaddy of modern urban-driven marketing systems. Where hand harvesting was once considered mean, back-breaking work, urban consumers, picking their own produce, view it as recreation.

Subscription gardening is one of the newer, urban-driven marketing systems. Customers, usually referred to as clients, pay a pre-determined amount of money to a farmer who in turn provides the client a season's supply of food. Clients might or might not help with the work. The farmer knows ahead of time what he needs to grow, and gets at least part of the subscription fee up front to finance the crop till harvest time instead of having to borrow it from the bank. I like this kind of arrangement because it engenders little farms right in towns and cities—hence, the birth of the "urban farm" as it is being called. If you could buy stock in something called "the urban farm movement" it would be a good investment. Fortunately, no such stock exists.

Whereas Sinclair and similar farmers sell mostly out of trucks or booths at farm markets or at their own stands in urban locations, other farmers sell very specialized high-quality produce directly to restaurants. Restaurant specialties might be baby vegetables, or gourmet mixes of salads, or exotic varieties of melons, and so on. I recently obtained from another gardener a golden raspberry, variety unknown, which is absolutely luscious, more resistant to mosaic diseases than the other yellow varieties I've tried, but only a fair producer. I wonder whether it would be of interest to local fine restaurants if I could supply only small amounts in season. Five years ago, a restaurant manager some distance away was not encouraging. "We can't handle local stuff, I don't care how good it is," he had said, "unless you can promise me year-round, weekly supply." He's not in business anymore either.

But now, when I asked Rosie Wood at our local Woody's restaurant, she responded most enthusiastically. Not only would she be interested in golden raspberries, but just about any other high-quality fruit or vegetable I could bring her. She has since sent me information about other farms from which she obtains vegetables, as a way of encouraging me. This is what I mean by an urban-driven agriculture. If it can happen here, in a county where the total population is less than 26,000, it can happen anywhere in America. The real problem is that we do not open possibilities like this to young people, either at home or in school. We persuade them to seek "secure" jobs with the likes of Dupont and IBM which are presently laying off workers by the thousands.

Another developing market for pesticide-free food comes from people who are learning that they suffer from chemical allergies hitherto not recognized as such. Allergists are discovering that some cleaning fluids, dyes, pesticides, and other chemicals thought to be "safe" cause, in some people, persistent allergies that have been diagnosed for years as flu or other common ailments.

A man came to me several years ago wanting to know if I would sell him unpasteurized milk from our cow. Doctor's orders. I had to tell him that the law forbids me. I can drink my own. I can give it away. But I can't sell it. Bootleg milk. Another example of idiocy done in the name of "serving the people." The real reason for laws against certified clean raw milk is to protect the established dairy industry's monopoly over milk production.

Keeping the Proving Ground Small

Whether cottage farms merely want to grow their own food, or decide to branch out into a commercial venture, the kitchen garden—the learning garden, so to speak—should be small. Almost everyone imbued with cottagitis, including me, initially makes a garden that is way too big for the time they have available to care for it. That is why some people get discouraged and quit gardening. On a small kitchen garden you can lavish compost and mulch. You can concentrate the time you do have available to keep weeds from ever gaining a foothold and so make gardening an easy job. You need only hand tools, thus cutting down significantly on the expense. You can handily irrigate, if necessary, and use only a little water. You can control bug damage, if any, by hand methods. The out-of-pocket cost per pound of food produced this way is extremely low compared to buying food.

My daughter and son-in-law have a raised bed garden only twelve-feet-square in size behind their suburban home, but the quantity and quality of food it produces is amazing. They fertilize with top grade, composted sewage sludge approved for gardening. Never anywhere have I seen such luxuriant tomato and squash plants. By using clever trellising and every square inch of ground space, and by putting in new plants as quickly as the old are harvested, they can raise all their spring, summer, and fall salad makings and most of their main-course fresh vegetables.

Keeping the kitchen garden small opens up other intriguing possibilities that make gardening enjoyable and recreational. My dream is to surround ours with a stone or brick wall as Scott and Helen Nearing did. The smaller the garden the more affordable a wall. A wall holds in heat and holds out thrashing country winds that folks in town seldom have to deal with—the concentration of houses there acts as a gigantic windbreak. Along with stopping the wind, a wall would radiate warmth absorbed from the sun during the day to the plants at night, and that could mean a week or two more of production at both ends of the season. Frost-shy fruits, like peaches, could be espaliered against the walls as the English do to take advantage of that radiated heat on those nights when a degree or two of coldness could make a difference in saving blossoms or buds from late frost.

More importantly, the wall would keep out rabbits and deer and, with a deep foundation, moles and groundhogs. Even squirrels, raccoons, and chipmunks would find the walls daunting. Since small gardens are so vulnerable to wild animals (a couple of rabbits or a groundhog can ruin a new planting in one night), barriers are important, and a rock or brick wall is beauty forever compared to the annual rigging of rabbit fence or electric fence that most of us have to use.

A wall would provide 100 percent privacy, too, so that on those warm spring days when you feel like shedding most or all of your clothes, you could do so without fear of offending neighbors. I have a theory that the purpose of those English walled gardens was mainly to create a safe haven for nude or almost-nude gardening. Those Victorians, as cultural history is proving, were a pretty wild bunch behind closed doors.

Keeping the garden simple is just as important as keeping it small when you have so many other things to do on the farm. Okay, so it's fun to try out something new and exotic every year. That's one of the functions of your "proving ground." This year we discovered a winter squash, *Delicata*, which we obtained from Pine Tree Gardens (Box 30, New Gloucester, Maine 04260), to be much better in texture and taste than any of the varieties we have been growing.

But you can overdo the testing of new and exotic plants and end up with a hodge-podge of everything, but not much of anything. Once you have found excellent varieties, quit growing the less satisfying. There will be no more Table Queen, Table King, Butternut, or any other conventional winter squashes in our garden. Also, just because a lettuce variety has a French name doesn't mean it is a bit tastier than Buttercrunch. Grand old heirloom varieties, another hustle, might have more distinctive flavor than new hybrids and then again they might not. Test them before going whole hog. ("Distinctive flavor" as used in many garden catalogs should be translated as "tastes funny.") Nor does an onion from Vidalia or anywhere else necessarily taste any better than one you grow in your own organically enriched backyard soil.

Simplicity has not come easy for me. I've had to learn the hard way that I might just as well stick baseball bats in my yard and expect them to root and grow baseballs as to plant apricot trees and expect to get apricots. The same with Green Gauge plums. We quit raising turnips, rutabagas, and parsnips because we finally admitted we didn't really like them. I got rid of all those onion things called nest onions, Egyptian onions, and potato onions which I had formerly been known to praise in print. Yes, for about a week very early in spring, they are fairly good but very soon the regular green onion scallions come along and they taste so much better that the wait is worth it.

The keep-it-simple rule applies even more, I think, if one decides to try to sell garden produce. In this case reliability is everything. Trial and error of thirty years in the garden has taught me that if I were to try retail sales, I should concentrate on asparagus, raspberries, strawberries, tomatoes, and sweet corn. These are the foods that by testing in the garden we know we can grow well and dependably on a larger scale.

Cottage farmers have some built-in advantages in keeping the kitchen garden simple. Because they have a relatively large amount of land compared to a city gardener, they can grow some of their food in other places on the farm. Sweet corn for example we usually treat as part of the farm operation, not the garden, especially in years when we want to grow a surplus to sell or give to other family members. Then I plant several very long rows along the edge of the field corn or in various plots around the farm, and care for them as part of the farm crop.

We generally view other crops that take up considerable room and which we plant in some quantity, as part of the farm, not the garden: pumpkins, cantaloupes, and watermelons to be exact. We generally plant these along the edge of the cornfield too. Farmers who follow new rotational pasture systems are planting turnips and kale for winter grazing and so obviously they could harvest some for themselves, too.

On a cottage farm, food from the wild also helps to keep the garden simple and small. Wild raspberries, blackberries, persimmons, pawpaws, hickory nuts, mushrooms, black walnuts, wild salad plants, and fish from farm ponds all figure in our diet and relieve some of the pressure to expand the kitchen garden. Perhaps the best examples of this kind of simplification are the many wild apple trees I have planted here and there around our farm. As it turns out, these trees are producing as much fruit, of as good a quality, as the commercial varieties I purchased for the formal orchard. We don't spray them, don't prune them, don't fertilize them. Just eat the fruit. By making apple production a function of the wilder side of the farm, we have simplified the formal orchard out of existence.

The Proving Ground is Our Common Ground

But taking into account all I have said, the most valuable transfer from garden to farm is cultural, not technological. It is in the garden that we get down on our hands and knees and feel the soil draw us into an understanding of the interrelationships between all living things. One generality that comes close to being always true in my experience is that farmers who do not garden or who have never gardened, tend to be insensitive to the biological nature of their work and therefore inattentive to all nature including human nature. Urbanites who do not garden are even worse in this regard since they have no frame of reference at all for coming to grips with the realities of biology. They not only don't understand what farmers are up against, but cannot see that these problems are everybody's concern.

On the other hand, the more gardeners immerse themselves in their biological art, the more they not only understand farmers but become farmers—nurturers of life. Indeed, no matter how small the garden, even as small as a miniature planting of mosses inside a gallon jar, the biological activity going on there is a microcosm of the farm.

It seems to me that the garden is the only practical way for urban societies to come in *close* contact with the basic realities of life, and if that contact is not close, it is not meaningful at all. To feel the searing heat as well as the comforting warmth of the sun, or to endure the dry wind as well as the soothing breeze; to pray for rain but not too much rain; to long for a spate of dry weather but not too long; to listen to the music of nature as well as the rock beat of human culture; to know that life depends on eating and being eaten; to accept the decay of death as the only way to achieve the resurrection of life; to realize that diversification of species, not multiplication within a species, is the responsibility of rational intelligence—nature will handle that latter activity much better than we can; to grow in personal simplicity while appreciating biological

50

complexity, so that in the garden there is time to sit and think, to produce good food for the mind—these are all part of an education that the industrial world hungers for but cannot name.

We humans are only beginning to understand the wondrous world in the soil beneath our feet. We still call this inner sanctum of life "dirt." How shall we change our view, how shall we learn how to insure the continuance of a biologically healthful earth now that humans gain the power, more and more, to destroy the earth? Where, but in the daily garden, can this education occur? The conventional farm in the embrace of industrialism can't teach this kind of reverence to society. Even in those instances where the farm could teach these lessons, it is too far away from most peoples' lives.

There are from 10,000 to 40,000 microbes in every teaspoon of soil, microbiologists estimate: tiny animals, tiny plants and fungi, many of them unnamed, many not even identified by a number. Dr. Harry Hoitink, at the Ohio State research center in Wooster, Ohio, who works in this world every day, can become almost lyrical in his descriptions of what he refers to sometimes as an "invisible but exotically beautiful jungle world." He has made some remarkable discoveries about this world: Various microbes in composted organic matter can control diseases in plants without fungicides. The world of soil microorganisms is the magma of life, and yet, as Dr. Richard Harwood at Michigan State University says, science does not have enough long-range data on what effect toxic pesticides have on soil biota. His experiments do show that the healthier and more diverse the microbes of this teeming world, the better plants grow. We need to aim powerful electron microscopes into soils across the nation and beam the sights and scenes of soil life by way of television to all schools, libraries, workplaces, homes, laboratories, churches, and governmental meeting rooms for society to view constantly, as it views the stock market reports and sport games. We could then watch this unseen life unfold and know who is winning, who is profiting, as trillions of living forms are engaged there in the sublimely feverish exchange of life and death. How quickly humankind's ideas on farming systems—and on life in general—would blossom with new understanding and wisdom.

I know of only two ways to move humans to become vitally interested in the very substance of life: By fascination or by starvation. Surely fascination is the better choice. Being thus entertained in the garden, urban and rural society would join hands in the preservation of nature and turn the earth into a garden of Eden.

CHAPTER 4

The Peaceable Kingdom of the Barnyard

What is man without the beasts? If all the beasts were gone, men would die from great loneliness of spirit, for whatever happens to the beasts also happens to the man.

Attributed to Chief Seattle, 1855

Innocent nature lovers think of wilderness as a place where they can go to quaff a little tranquillity—as if tranquillity could be bottled and sold like spring water. They are able to coddle such a fancy only because civilization, which they in certain moods pretend to loathe, allows them to "experience" nature's seeming peacefulness without having to come to grips with its unrelenting violence. They do not understand that Mother Nature can just as often be Old Bitch Nature. They drive a van into the mountains or fly a plane onto a secluded lake, taking along enough industrial luxuries to keep them comfortable for a week or two. They sip the wilderness briefly. They shoot off guns. Roar motors. Guzzle beer. Play cards. Click cameras. Life doesn't get any better than this, but they hurry back to civilization as soon as they run out of food and film.

Nature is a vast killing field. No bug, plant, or animal including humans can live unless other bugs, plants, or animals die. All we do is trade corporeal forms around the gaming table of existential matter.

Wild animals live with one eye over their shoulders watching for predators, and the other eye looking ahead for prey. I listen to a red-winged blackbird, warbling his sweet song from a bush above the creek. How contented he sounds. In reality, ornithologists believe that the translation of that sweet song goes something like this: "This is my territory and if I catch any other redwing trying to move in here, I will peck his beady little eyes out." And that may be as close to contentment as a songbird ever comes.

Domestic animals enjoy much more tranquillity than wild animals. My two hogs live a life of comparative luxury, snuggled in their straw, deigning to roll out occasionally to eat the food I put before them. My cow lies languidly in the shade while I sweat in the hay field. Her only real stress in life are the flies that in July swarm around her.

If these animals are slaughtered for meat in the end, so too do the wild ones die. Watch an owl tear the guts out of a rabbit. Watch a black rat snake swallow a ground sparrow. Watch a fox pounce on a mouse. Watch wolves hamstring the aging white-tailed buck, too old to run away. Which death is worse, really: the old buck, half starved, stiffened with arthritis, torn slowly to pieces by wolf fangs, or the buck in its prime, shot by a man? In the wilderness the only unnatural death is a natural death.

Even the barnyard can be dangerous ground. If I were to faint in my pig pen, these seemingly lovable porkers would eat me. Hundreds of children on the farms of yesterday bore witness to this fact, and distraught parents learned of the tragedy by spying a pig running across the barnyard with a bloody arm or leg in its mouth. Life is *by nature* dangerous. That is the first lesson an education should teach, but education does not teach it, and as a result we have a modern human society insisting on a zero risk environment that cannot be.

To maintain a modicum of tranquillity in the barnyard, there are three cardinal principles that I try to follow. The first is to realize that animals do not like change. I establish a routine with them and stick to it as much as possible.

Secondly, in teaching them a routine, or a necessary change in that routine, I don't try to force the issue. I bribe them with food. For example, until she gets used to the idea, a new heifer is afraid to stick her neck through the neckhold of her stanchion. Trying to force her only causes mayhem. I put grain in her feed box in such a way that she must put her head through the stanchion in order to reach it. The first few times she obliges, I do not try to close the stanchion against her neck. About the third day, while she gobbles away, I can slip the neckhold closed and she hardly notices.

54

Thirdly, I have learned not to try to handle just one animal alone unless the animal is accustomed to being alone. A flock of sheep want to be together and an individual sheep will, like a teenager, get very upset if separated from its crowd. In a pen alone, it will panic when I try to catch it and hurt itself (or me) trying to jump out. When I need to catch a couple of ewes to trim their hooves, I coax all the sheep in the barn with food and let a bunch of them jam into one pen. Then I can push my way gently through them to catch very easily the one I want. Likewise we can worm the lot of them while they stand placidly together. When shearing, bunch the sheep as close to the shearing platform as possible. That way, you don't have to chase them, and it means less distance to move the animal to be shorn after you catch it. Then it will remain quieter during shearing because it can see the other sheep close by.

Kindness to animals is well worth the effort, but never take for granted that the animals will be kind to each other. Put a strange hen in with your flock and the resident chickens will peck her to a bloody pulp. A boss cow will butt cows lower in the herd pecking order away from her self-ordained place at the feed bunk. If there is not enough room for all, the cow at the bottom of the social order may not get enough to eat. If cows have horns, the social order problem can get gory, pun notwithstanding. The humane way of removing horns is to do the job when the horn buds first protrude on the calves' head. The dehorning paste will then burn away the bud with the least amount of pain. A better solution would be to breed the polled (hornless) characteristic into all cattle or at least, as an individual cottager, to raise only polled breeds.

Despite these nastier tendencies of animals, our barnyard is a much more peaceable kingdom than either the wilderness or the bleak concrete and steel-cage realm of the animal factories. Needless to say, it is also far more peaceable than the public haunts of humans where senseless violence unconnected to the food chain of eating and being eaten threatens to reduce "civilization" to hellish chaos. My hens are in a little danger from hawk and Great Horned owl, fox and coyote, but not nearly as much as I risk merely driving on a public road on a Friday night when the drunken drivers are weaving their way homeward.

The hens wander through barnyard and field where food can be found at every turn, and then go to roost in a safe coop at night. They

55

sing as they stroll on their daily rounds, a sound that recalls to me the time when every barnyard was full of animals and every farmhouse full of farmers and we were all so full of food and comfort that we could not believe the news—that people in New York City were standing in bread lines. Why didn't they go to the country and get a piece of land, Grand-paw would keep asking. It seemed so simple to him, secure in his barn-yard with centuries of survival music to assure him: hens clucking, hogs squealing, cattle lowing, sheep blatting, roosters crowing, horses whin-nying, bees buzzing, calves bawling, sons arguing, daughters giggling, and Grandmaw calling him in to dinner. If we lived such a dull life compared to our "urban counterparts," as the sociologists (the sons and daughters of those breadlines) say we did, why was my family always singing?

Having only a few animals, and getting to know them well, I fall into the habit of talking to them. If I discourse on the inevitability of death, they, of course, do not understand. At least I don't think so. But they understand plenty about subjects of interest to them, like food and security. When I am carrying the bucket we reserve for table scraps, the hen flock will shun their normal milled grain and follow me all over the barnyard. They love table scraps above all other morsels. I think of the millions and millions of tons of table scraps that go by way of garbage disposals into the sewer systems of America. For our wastefulness, we deserve to stand in breadlines.

If I pitch my voice into a high tremulous wail, mimicking the faint siren-like call that the hens use themselves to signal danger, young chicks will dash to their coop. If I mimic that same sound to an egg that is about to hatch, the chick within will stop peeping. If I cluck like mother hen normally does to signal all is well, it will start peeping again.

My animals notice more about me than I do. They can tell when I am happy, sad, angry, amused, impatient, or too tired to care. They know especially when I'm in a hurry to get the chores done and when I am in a loafing mood. In the latter case, they stand in my way, butt me, rub against me, stick their heads up to be scratched. In the former case, they stay out of the way and reflect the same nervous tension that I am unintentionally showing. Only by studied concentration have I learned to move slowly among them, and to confront them first with my nose, as they do me, not with my hands. Reaching out at an animal not yet wholly tamed, will frighten it. Perhaps because they have none, animals are edgy about hands.

They even notice changes in my appearance. One morning the chickens acted like idiots when I entered the coop, flapping themselves into the far corners. "What in the world is the matter with you?" I asked. They continued to cower even while I poured wheat into their trough. Then I remembered. I had on a different jacket and hat than I usually wear.

And how jealous of each other animals can be, especially the bovines. Our cow will butt the smaller steer away if I scratch his ears. I am supposed to scratch her ears.

I can distinguish different moods in the animals too. A sick cow gets an opaque glaze over her eyes that is unmistakable. A cow in heat is possessed of a nervous intensity that alters her usual calm personality dramatically. I can tell the change in our cow the moment I enter the barn, whether or not she is trying to ride the steer or one of the sheep. The drive to reproduce is a powerful force. A pregnant cow or a cow with a calf at her side is far more contented than one with an empty womb. Similarly, a ram penned away from the ewe flock bellows and raves and runs back and forth along the fence, reminding me of male fans at professional football games. When the ram is with the flock, he settles down, full of contentment, even during the many months when there is no mating, and so no "outlet" for his assumed sexual energy. Do animals know something we don't?

Animals are more literate than humans in reading body language. If I try to stop them from going where they've made up their minds to go, they can anticipate the direction of my lunges before I lunge. A human possessed of this skill would make the greatest basketball or football player of all time. Our pigs know if they try to nip the back of my leg, that I will slap them, but they see the slap coming before I execute it and squeal mischievously as they easily avoid it. I think leg-nipping is a game with them. Ewes warn intruders away from their lambs by stamping their feet. Turned out on pasture that first day of spring grazing, the calves articulate their delight by kicking up their heels and the lambs by bouncing stiff-legged across the pasture. When they have eaten their fill of grass on a breezy summer day, sheep lay down on the brow of a hill where they can catch the full force of the wind to blow the flies away and snooze with their heads tilted up in the air, their faces silently telling the world how happy they are.

The ultimate example of body language literacy takes place between a flock of sheep and a sheep dog. The first time I saw a dog load a flock of sheep I was convinced the sheep had been trained to hop up in the truck whenever the dog came into sight. To the human eye, the body language between the sheep and dog is not fully observable. Dog and sheep anticipate the movements of each other before these movements actually occur, giving the impression that the sheep understand beforehand what they must do when the dog approaches. I know of one dog so talented that it will drive a whole flock of sheep to a fence corner and hold them there for hours, while the shearers remove their fleeces one after another. I have watched a shepherd turn his dog loose in a large valley pasture, and by certain whistling commands, direct the dog to bring an entire flock right up around him. That dog could even single out any specific ewe and hold her by eye contact next to a fence for the shepherd to catch and examine.

Animals have distinct personalities. Some are friendlier than others, some much less trusting, some more adventurous or curious than others. One Plymouth Rock pullet likes to sidle up to me and peck me on the leg ever so gently. I have no idea why. None of the others do that. No chicken I ever raised before did that. Another, a Rhode Island Red hen, always accompanies me as I fork manure into the spreader. She eats bugs and worms I uncover with the fork. The other hens wait until I leave before moving in to join her.

The animals likewise have different food preferences: two ewes show little interest in even the best hay if there is oats to eat; others given a choice of oats when good hay is available react to the oats the way a child reacts to broccoli when candy is the other choice. The steer I am raising for our next year's beef supply would not at first eat his milled feed unless I dribbled a little oats on it and he definitely preferred high quality clover hay to any kind of grain.

Whenever I switch from one mixture of milled feed to another, our cow is furious because the new stuff tastes differently. She will sniff at it, shake her head, stare at me malevolently, knock her feed bucket around with her nose. In four different ways she is distinctly telling me that she doesn't like this new stuff.

"That's all there is," I say, running my hands through the feed and licking my lips to show how good it really is if she just takes a mouthful.

Again she shakes her head and stares at me so irritably that I have to laugh. That makes her angrier yet and she shoves the bucket away with her nose. "So there," she is saying, plain as day. "Take this crap back to the kitchen."

Not until I walk into another part of the barn and she realizes that I am not going to give her anything else, does she sample the new feed. Sniffs disgustedly and shakes her head again. Finally takes a little mouthful. Chews. "Hmmm, not so bad after all." Then having come to terms with the new smell of the feed, she dives in and gobbles it all down. If the younger steer even thinks about approaching, she gives a signal I cannot catch, and he backs away to his own stanchion and waits patiently for his own ground corn—topped with whole oats of course.

My mother once said that she preferred the company of cows to that of most humans. I thought she was joking, but as I grow older, I'm not so sure. The intellectual stimulation of conversing with animals is very low, granted, but there are other satisfactions especially for the husband-man whose other job batters him with long hours of human babble. Farm animals, if not hungry or separated from their usual companions, are *quiet*.

There is a deep satisfaction in scattering clean yellow straw knee deep for the animals to sleep on and then feeding them in the still of a wintry eve. Sheep give the most contented little sighs when they nose into their food. Horses snuffle in their hay, and the soft munching sound of cows chewing their cuds rises serenely to the hay mow where I sit and listen. The mother ewe with coaxing grunts encourages the new lamb to nurse and finally the smacking sound of the lamb sucking vigor-ously reaches my ears. All is well. It is no surprise to me that a god might choose a stable to be born in; only the ignorant think such a birthplace would be below a god's dignity.

A summer day in the barnyard is equally as tranquil except for the pesky flies. Trees cast shade over the henhouse and the pig house, blunting the swelter of the sun. A breeze sifts through the open barn doors, cooling me as I pitch forkfuls of manure into the spreader. My little red hen stands practically between my legs, head cocked for worms and fly eggs that my forking turns up. The other hens fluff their feathers contentedly in the dust holes they have made near the machine shed, stirring so much air into the dirt that it has, in its fine, silken-silty dryness, almost the quality of water. Human expertise believes the hens dust themselves to avoid lice. I think dust bathing just feels good to them.

The pigs awake from their almost continuous snooze, and while one of them plays with a piece of wood I have put in the pen for that purpose, the other tries to hunker into its water trough for relief from the heat. Pigs like nothing more than to wallow in mud, and I am perhaps a little insensitive to pig natures by keeping ours imprisoned in a pen raised two feet off the ground with slatted planks for a floor (see page 76). But it is either that or put rings in their noses to keep them from rooting under the fence of a lot, or rooting up the pasture or wrecking the fragile balance of life in the creek by churning it into a muddy mess.

The sheep and cow have come from the hot pasture into the woods, where they lay or stand in the shade, shaking heads, stamping feet, swishing tails, trying to avoid the eternal flies. They watch me closely from a distance. If I would open the door that gives them access to the barn, they would quickly pile inside, where the cool darkness discourages flies better than any fly spray. But I am not about to let them in now—bothersome enough stumbling over a chicken as I fork.

The two cats, my defense against rats and mice, have crawled up on the fender of the tractor, which is hitched to the spreader. They think they are very clever to have found a new place from which to watch me work. One of them lunges playfully at a horsefly buzzing by, misses, and unused to the hard, slippery surface of metal, slips and topples to the ground. Embarrassed, she walks away, pretending, I think, that she fell on purpose.

The only time the barnyard is not pleasant is in March. Mudtime. Then I am glad that I positioned the barn precisely at the highest location of the woodlot, so that water drains naturally away from it. Nevertheless the newly thawed ground around the barn is mushy. I keep the animals inside or they would churn their lot into a buffalo wallow.

Where I walk my daily rounds, I lay down old boards so I don't have to slip and slosh in the mud. At the entrance to the barn that the animals use, I have put down gravel to fight the mud. But coming at the end of winter, mudtime is endurable. Warm weather is on the way, and the ground will soon turn solid and dry again.

Many books present complete manuals on how to raise domestic animals. I will give here only the kind of details from our own experiences that I don't much see in books.

Chickens: The First Choice of the Cottager

Chickens are the easiest farm animal to raise and the most economical. Some say their meat is one of the more healthful to eat. But if you don't eat meat and don't like to kill animals, chickens are still appropriate if you do eat eggs. You can let the hen live out her time as an egg layer (five years or more), and then bury her in the garden for fertilizer.

If you don't eat eggs either (believing in the modern superstition that they contain too much cholesterol), you can raise exotic fowl for the pet market. I sometimes wonder why I mess around with chickens worth only a couple of bucks apiece when I could raise emus worth a couple hundred bucks apiece.

There are literally hundreds of different kinds of fowl amenable to the cottager's barnyard, from ordinary Rhode Island Reds to exotic golden pheasants, from homing pigeons to peacocks, from guineas to turkeys, from ducks to geese. A neighbor (I use the word neighbor to refer to any cottager in our county that I know well) makes his living raising thousands of quail which he mostly sells to nature or hunting preserves. Catering to poultry fanciers with exotic fowl can be very profitable and is an example of the offbeat approach to farming that is characteristic of the inventive cottager.

But the main reasons that poultry should be the cottager's first choice are these: they convert feed to meat more efficiently than the other conventional farm animals, and they will eat all kinds of waste food except citrus rinds; furthermore, unlike cows and hogs, anybody can catch and carry a chicken, so handling and transporting them is not a problem as it is with large animals. To catch a chicken all you need is a six foot length of stiff wire with a tight crook at one end to hook its leg with.

Generally speaking, a broiler, a chicken raised for meat, will gain a pound for every two and a half pounds of feed, a little faster gain if the feed is mostly corn with a high protein (soybean meal) supplement added. A full grown layer will eat about a fourth of a pound of feed a day, maybe a little more, depending on its size and the nutrient value of what it eats. That equates roughly to ninety pounds or a bushel and a half of corn per year. An acre of corn at 140 bushels per acre would accordingly, rear ninety chickens. A free range hen would actually eat less corn than that because of other food she forages. The small Bantam breed will eat half that. In fact a banty running loose seldom needs to be fed at all. In most cottage farm situations it will find its own food.

My laying hens get about half their food from bugs, grass, weeds, and worms as they range over field and woods, and from eating the grain that falls through the slatted floor of the hog pen, and what the cows and sheep dribble out of their feed bunks. In fact our chickens are our in-house garbage collection service, scavenging up all our table scraps plus every bit of stray grain that might otherwise draw rats. One of the neatest ideas I've heard recently is to move chickens onto pasture plots behind sheep or cattle in a controlled grazing system. The chickens scratch and scatter the manure of the larger animals as they eat the bugs and worms drawn to the manure. They also eat the eggs of sheep and cattle worms, thus helping to solve internal parasite problems. The chickens can be kept in movable coops, and the whole coop moved from paddock to paddock. Chickens always (almost always) go back into their coop to roost as darkness falls, so there is no problem moving them.

You can raise four hens in the backyard easier than you can keep a dog, although because of our weird zoning ordinances, dogs are okay and oftentimes hens are not. The only "secret" I know is to provide a lot more room for your chickens than the 1.5 square feet per bird that many expert manuals call for. Allow at least five times that much room per bird, and better ten times that amount. Space requirements in commercial poultry houses have to be kept at an absolute minimum because of the almighty dollar, but not in a cottage farm environment. Our 10-by 20-foot henhouse is perfect for twenty hens, although for two months in the summer we also fatten thirty broilers in half of it. The somewhat crowded conditions at this time do not much matter because the chickens are outside most of the summertime anyway.

Sure, the cost per square foot of space is higher in an uncrowded coop but I am sure that when all the expenses are figured in, the cottager will eventually gain back the initial difference by subsequent savings. In a commercial poultry house, the cages that make such an "efficient" use of space are not cheap. Water and feed and medicines have to be delivered to the cages by costly automated means, expensive air circulation systems have to be installed, and getting rid of the manure requires tremendous investment in handling equipment. All this is unnecessary in a cottage coop where chickens roam outdoors much of the time. In an uncrowded coop, you can nail a two-by-four about two feet high across adjoining walls to provide plenty of roost for twenty to thirty hens. With two feet of sawdust or straw for bedding, the chickens by their constant scratching turn their manure into the bedding and transform both into a wonderful, granular compost dry and odorless enough to handle with bare hands. The chickens eat tiny specks of humus containing Vitamin E from this compost, thereby decreasing significantly the incidences of cannibalism. Mine have never become cannibalistic. I have never fed them any medicine of any kind for twenty-seven years.

A crowded coop becomes sodden with stinking manure. The chickens get manure on their feet and get it on the eggs when they go into the nest to lay and then you have to wash the eggs. Ideally eggs should not be washed (the shells are quite porous) and they rarely need washing in a coop with lots of room for the hens and with nests routinely replenished with clean straw. You will counter the costs of years and years of uncrowded poultry production in this manner by avoiding just one case of salmonella.

It is a good idea to divide your coop into two parts with a door between and an entrance/exit door to both sections. Such a division allows us to keep our young chicks separate from old hens until they are of laying age. We built three nests along one wall, enough for fifteen to twenty hens, each nest about fifteen inches square and roofed over so that the inside of the nest stays relatively dark. Darkness discourages egg eating. The roof over the nests should be very steep so the hens can't roost on it and cover it with manure.

In addition to our own whole corn and wheat, we feed the chickens oyster shells which are supposed to keep the eggshells strong and provide a little grit to help them digest their food. They can get more grit (tiny pieces of stone and sand) from the soil. Sometimes I feed milled corn to the broilers to make them grow fat faster, but the layers receive only whole grains. They all get fresh water daily. I put the water in receptacles made from the bottom halves of plastic detergent containers. These containers don't crack when water freezes in them, and by knocking them against a tree or wall you can break the ice out easily. Rubber containers also work well this way.

There is no electricity at our barn, no running water, and no water heaters. The water for the animals runs off the barn roof into two barrels, one of which is in-ground and covered with wooden boards and overlaid with an old piece of carpet and snow when available so water doesn't freeze in it in winter. I dip out water as needed. The cows and sheep usually go to the creek to drink, or eat snow. Very primitive. No pipes to freeze, no motors to burn out. Without "labor saving" technological gadgets to help me, I save a lot of time by not having to fix them.

We buy chicks in May after the weather has warmed up enough so we do not need heated brooder houses. The chicks stay in a large cardboard box in the garage the first week, with an electric bulb above them for a little extra heat, if necessary. Then they go to the coop, but separated from the old hens. We feed them ground feed until they are old enough to eat whole grains. The heavy White Mountain Cross broilers, bred to gain weight fast, eat much more than the finer-boned laying breeds and are ready to butcher in about two months. The layers are then gradually introduced into the layer flock and the door between the two parts of the coop is then left open. Three to five year old hens no longer laying we butcher for chicken soup.

Sheep, the Second Best Choice for the Cottager

Just as chickens provide three products, meat, eggs, and compost, so do sheep: wool, meat, and an excellent manure for fertilizer. Because diversity is such an important key to cottager efficiency, this triple market potential is something to value. The wool payment can sometimes pay for the ewe's purchased feed, and the manure will replace most or all of the purchased fertilizer that would otherwise be needed, so the money brought in from lambs fed out on pasture and mothers' milk, is almost all net profit minus the labor involved. Last but not least (by far), even older people can handle sheep for hoof trimming or shearing or lifting into the truck if necessary. Try that with a steer or market hog.

Because of their triple value, sheep are a good choice for a vegetarian farmer. They can be kept profitably for wool only, especially where one can find a specialty market with a higher price than the common commercial market or where the cottager is a spinner or weaver and sells the wool as yarn, fabric, or finished wool products.

Sheep have a fourth function on the farm as scavengers. They will clean up weeds in fencerows, and eat grain from the fields missed by the combine.

Except during shearing and lambing season, sheep do not require much attention and so are ideal for a cottager with limited time. Shearing and lambing take place here in late March/early April when there is no other pressing work on the farm. From May until winter, the flock mostly takes care of itself on pasture. I say that with my fingers crossed, because, as with everything else, the more attention you pay to sheep, the better they will do. For example I believe the reason I have thus far escaped coyote damage is that I make it a point to walk through the pasture almost every evening, and when the lambs are small, every night, shining a flashlight up and down the valley (coyotes don't like bright lights) and even shooting off my rifle if I hear howling in the distance. I also lure the flock up close to the barn at night with some choice morsel of hay, grain, or fruit.

The general practice among shepherds in humid regions like Ohio is to cut the tails off the lambs about a week after birth. (In the dry western range country, this practice is rarely followed.) Most often this is done by slipping a very tight rubber band (elastorator) around the tail with a special tool. The elastorator cuts off the blood circulation and the tail drops off in a few days, presumably causing the poor lamb only minor discomfort. I do not like to "dock" lambs at all, and this way least of all. On lambs I intend to keep for ewes, I cut the tails off cleanly with a hot nipper tool, cauterizing the wound in the process, after applying a tourniquet (which everyone tells me is unnecessary) and then applying pine tar or a medicated wound powder. I leave an inch or so of tail, which heals faster than the show ring practice of amputating clear back to the rump.

Most shepherds feel obliged to dock lambs because the stockyards will generally knock a few dollars off the value of every lamb with a tail, no matter how clean and healthy the lamb. I have argued at the stockyards, to no avail. The stated reason for docking is that manure will catch and build up on the tail. The gob of manure may become infested with maggots which then eat into the lamb's flesh, certainly a fate worse than docking. But I fatten my lambs on pasture—no grain—and only one year did a few lambs have a manure build-up on their tails, during an autumn of excessive rains that made the pastures too lush. Even then the problem did not result in maggots.

Last year I found an auction market where I can sell direct to buyers, with the stockyards taking a commission. The buyers last fall paid just as much for tailed lambs as for tail-less ones, everything else being equal, so I will no longer cut off tails on market lambs.

Ram lambs to be sold as market lambs are generally castrated, although it is arguable whether meat quality is improved by the practice, so long as the lamb is sold within about half a year of age. Iowa farmer and magazine publisher Maury Telleen has kept sheep all his life and he tells me that he does not castrate market lambs anymore. But a $5 per head penalty may be levied against uncastrated fat lambs when they are sold, so most shepherds feel obliged to do it. Generally castration is done by putting an elastorator around the scrotum, which, like the tail submitted to this method, eventually sloughs off. I would worry about infection, not to mention unnecessary cruelty. Instead I have the vet neuter the animals with a special instrument, called an emasculator castrator or burdizzo, that severs the seminal ducts just above the scrotum, a bloodless, safer, and more humane method similar to human male vasectomy. At the same time, the vet gives the lambs a shot to protect them from overeating disease. (The medicine is *Clostridium perfringens* Type D Toxoid. Sometimes the digestive tracts of lambs can't handle a big intake of fresh grass, and the normal activity of microorganisms in the gut breaks down, resulting in a sudden buildup of one kind, *Clostridium perfringens*, which can lead to death.) Then the flock goes out on pasture and except for a possible mid-summer worming, it requires no more work from me other than shifting the sheep from one pasture to another. I don't wean the lambs. They nurse until the ewes dry up which means the lambs are still getting milk when they are quite large. I think this is one reason that I can get hundred-pound lambs without grain. If you handle a flock this way, not weaning and separating lambs from their mothers as most people do, then you must castrate or the grown lambs will breed the ewes.

There is yet another way around almost all the gruesome work of tail docking and castration: Easter lambs. There is an ethnic market at Easter time for lambs of about six weeks of age weighing under about thirty pounds—the equivalent of veal in bovines. These lambs require no surgical alterations, and because the price is usually about twice that of conventional fat lambs, and since they require little or no feed or labor, raising them can be profitable.

Some sheep breeds are raised only for wool, like the fine-wooled Rambouillets, and some mostly for meat, like Suffolks. Most breeds raised primarily for meat produce wool that may be low in quality but still salable for some purposes. Many breeds, like Corriedale and Columbia, produce both meat and fairly high quality wool. Our sheep are mixed breed, but mainly Corriedale. I chose this breed for the very practical reason that we live near the Kin family which raises nationally renowned Corriedales. I can rent a buck from them handily every fall for breeding.

The tendency among sheep breeders in this part of the country is to breed for larger and larger sheep, be they Corriedale, Columbia, or Suffolk, and try for a 120-pound market lamb. As I grow older, I prefer smaller breeds like Dorset, Cheviot, and Southdown simply because they are easier to handle and I'm not convinced that two 120-pound lambs are more profitable than three 80-pounders. Arguing the merits of one breed over another is an endless sport and leads to one conclusion: raise the breed that you like best.

Proponents of newer breeds which routinely birth three, four, or five lambs at a time, including Finnsheep, Polypays and others, would have us believe these sheep are more lucrative. This is arguable too. In most cottage situations, I think you will find sheep a more enjoyable and even more profitable enterprise if you can improve your flock by selecting out only ewes of whatever breed that generally birth twins, with now and then a set of triplets to make up for the singles. Some super-shepherds, who concentrate all their time and effort into sheep and nothing else, can disprove that contention by averaging three lambs or more per ewe with certain breeds, but many shepherds I talk to agree with me—averaging two lambs per ewe is enough. Why do we farmers constantly fall for the mistake that we will get ahead if we out-produce each other? All that does is drive the price down. If we had ewes that could raise a dozen lambs each, we would be no better off as a rural society than having ewes that will raise two lambs each.

By the same token, specialists in the sheep business are going to twice-a-year lambing. That practice doesn't fit the cottage farm regimen as I see it, either, since the fall lambs will require expensive hay and grain. The principle I follow here flows from the same economic philosophy I have mentioned before: When you get to be super-duper, you also have to put in super-duper time and expenses and if nature doesn't cooperate, maybe endure super-duper losses. If you need more money, instead of gunning for two lambing seasons a year, do one lambing well, and raise something else in the fall instead of a second batch of lambs.

Lambing time requires almost constant attention. The shepherd must be there if midwifery is necessary. That means checking the barn every few hours. Lambing on cottage farms ought always to take place after cold weather has passed, as nothing will kill a feeble lamb at birth faster than losing body temperature. On the other hand, a healthy lamb with a normal birth can tolerate surprisingly cold temperatures. Lambing on spring pasture is the best alternative, I think, if the threat of predation from dogs or coyotes is minimal.

Only experience can teach you how to manage lambing. First of all, subscribe to *The Shepherd* and *Sheep* magazines and study them for a year. Beginners are usually too solicitous and try to get the newly born lamb to suckle right after it is born. Instead, after you get mother and lamb isolated in a lambing pen, *leave them alone for a while.* Often a ewe will back away from a lamb the first few times it wobbles up, searching for her udder, and I have become convinced that she does so simply because I am watching and making her nervous. So I go away. If the lamb is crying loudly an hour later, I come back and see what's the matter. Tip mother up on her butt, and check her teats to make sure they are not blocked. Although I have never yet been able to force a lamb to suck, sometimes if you squirt some milk on its nose, it will get the idea. Almost always if it will not drink, the ewe is butting it away when you are not present or there is something internally wrong with the lamb. When a ewe will not accept a lamb, the reason usually is that she has twins or triplets, and the first lamb born wanders off while she is in labor with the others. An hour later, when she is busy nursing the second lamb, she does not recognize the smell of the first one and refuses to accept it. Smearing some afterbirth on the unclaimed lamb's rump sometimes works (ewes recognize lambs by smell, and always smell the lambs at the anus). Just keeping both lambs together for a while sometimes works. Milking some colostrum from the ewe and feeding it by bottle and rubber nipple is the last resort. If the lamb gets weak and cold, take it to the house to warm it, and feed it mother's colostrum then. We have saved an almost dead lamb by submersing it in warm water for a while, then wiping it dry and putting the lamb in a box next to the stove. Every shepherd should have a woodburning stove (but a hair dryer will do). Sometimes recovery is remarkable. The lamb will wake up a half hour later, baa-ing for milk and trying to jump out of the box. If a lamb is obviously sick or disabled at birth and early attempts to help it draw no response, I do not persist in trying to save it. Rarely do extraordinary measures pay off. Let the poor thing die.

Sometimes you may end up with a lamb no ewe will claim, as we do occasionally. We put Pampers on ours, feed them by bottle, and let them have the run of the house. Fortunately, we've never had more than one at a time. Our "house" lambs snuggle in our laps and seem as much lulled to sleep by television as I am.

In about a week, a house lamb should go back to the flock. If it bonds too closely with humans, it will be extremely unhappy in the barn. Pet lambs have been known to pine away and starve from being separated from their human "parents." "Bounce," whom we rocked like a baby, did finally get used to being a sheep again and today she is very handy when I want to lead the sheep across the creek or into a new field. She will go anywhere I go except up a tree, and then the rest will follow.

Sheep are prone to internal parasites mostly because humans insist on trying to raise too many of them per acre available. As my neighbor, Al Kin, a nationally known Corriedale breeder, often says: "Shepherds begin with a small flock and have good luck. So they keep increasing flock size until they get in trouble."

How many sheep are too many? An acre here is supposed to carry five sheep through a grazing year. But if parasites are a problem and are becoming immune to wormers in your area, two sheep and their lambs are better, I think. If you are aiming at more profit, add one small-breed steer or cow along with the two sheep and their lambs per acre. Cattle parasites don't infest sheep and vice versa. If your pastures improve enough to carry more than that, increase only cautiously. We used to think that rotating sheep on pasture so that a given field or paddock was left ungrazed for thirty days broke the worm cycle, but animal scientists have proven that to be an error. A good cold winter will kill most worm eggs in the soil, and thus break the worm cycle temporarily. But even after a hard winter, some worms remain and there may be a rapid buildup in summer on crowded pastures.

Temporary pastures, rotated to other crops, pose less danger from parasites not only because the plowing and cultivating tends to clean infested soil, but because at least one year in four, when the field is in corn, it is almost entirely free of sheep until fall. That will break the worm cycle. The disadvantage of temporary pastures is that they can't withstand much animal traffic in wet weather and they do not contain a wide variety of grazing plants as a permanent pasture does.

We will carry sixteen ewes and their lambs and a couple beef cattle on about ten acres of permanent pasture and six acres of temporary pasture this coming summer. I hope to get thirty lambs from the ewes, to have forty-six sheep altogether and feed them exclusively on pasture from April through November when we sell the lambs. Only experience will tell for sure, but I believe that forty something will be our limit in sheep, not in terms of carrying capacity but because of internal parasite pressure. Additional grazing animals will be bovines, along with maybe a llama or donkey to guard the flock from dogs if that becomes necessary.

When lamb prices are good, a situation we are fortunate enough to be enjoying at the moment (late 1992 and early 1993), sheep can provide a very respectable income per acre without overcrowding. At a price of about 70¢ a pound, thirty lambs will fetch about $2000. (It is riskier to count lambs before lambing than chickens before hatching.) If the wool payment covers any extra hay I might have to buy for the ewes in winter, that $2000 means a net of about $120 per acre (16 acres involved). That meets my long-term goal even before adding in income from chickens, beef, and pigs, which are also fed mostly from these acres. That would seem to demonstrate that a smaller commercial farm of 160 to 200 acres, managed in this traditional cottage way, would make an adequate living for an economically conservative family. And remember that I am usually taking the easy market. If I market lambs directly through sales of meat to consumers and wool to spinners, as I sometimes do, a better return per acre could be achieved. A friend in Iowa tells me that he can made about $120 per lamb through private locker sales verses $85 through regular market channels.

A New Kind of Family Cow for the Cottager

The term "family cow" refers traditionally to a cow kept for a family's own milk supply. Such a cow will return more money to the cottager than any other animal on the farm, but also requires the greatest amount of labor including every-day milking. Realistically, not many families are going to want to, or be able to, milk a cow twice a day, nine to ten months out of the year.

I propose rather a cow kept primarily for baby beef—for the meat her half-grown calf will provide—and milk only if the cottager so desires. This alternative allows two options: Either let the calf take all the milk, or milk enough for the table and let the calf have the rest. The latter choice, which I have made, allows you the freedom to milk once a day, or once every other day, or maybe only once every third day. If the calf is born in spring, it can be ready to butcher as baby beef by winter having been fed only milk and grass and none of the precious supply of winter hay.

These options require a cow that gives only a medium amount of milk when fresh. A heavy producer will give more milk than a calf can handle and you will have to milk every day whether you want to or not.

71

(As an alternative, you can buy an extra calf and persuade mother to adopt it along with her own calf.) The disadvantage of a moderate producer is that by half way through her lactation, she may be giving enough milk only for the calf. However, since the main goal in this enterprise is the meat and not the milk, this is hardly a disadvantage for the cottager who is not keen on milking anyway.

A spring-born calf, raised entirely on plenty of milk and good pasture, reaches about 600 to 700 pounds by fall. Having never been weaned, its meat is luscious and tender: more tasty than veal but not as fat-marbled as beef finished to over a thousand pounds on a heavy corn diet. To our palates, baby beef is the best beef of all, and according to tests that my Kansas rancher friend Oren Long ran on his baby beef, as low in cholesterol as chicken.

In my opinion, the best cows for raising baby beef are Jerseys bred to Angus bulls. Mother cows that are crosses between Jersey and Angus, or Guernsey and Hereford are also ideal. These breeds usually give more milk than beef cows, but throw calves that are chunkier than dairy types. When taking some of the milk for table use, I have found that I had to milk at the same time the calf nursed—I work the two teats on one side of the cow while the calf is sucking the two on the other side. If I try to milk without the calf nursing, the cow will not let her milk down, as we say, saving it instead for the calf. Animals aren't stupid. That also means that I have to pen the calf for at least five hours before I intend to milk out my share. The rest of the time, the calf runs with the cow. It is much better for mother and offspring to be together has much as possible. I am convinced that calves are so prone to scours (diarrhea) these days because of the stress of being separated from their mothers right after birth.

It hardly pays to keep a bull for only one or two cows, so the cottager must generally turn to artificial insemination if a neighbor doesn't have a suitable bull. Artificial insemination is a great alternative but it does have drawbacks. Cows don't always "settle" on the first service and sometimes not on the second, third, or fourth either. This can throw your schedule way out of whack especially when you want calves birthing in spring to take advantage of a full season of pasture. Also some areas may not have an A-I service person at your beck and call. As dairy farms get larger, dairymen have learned to do their own breeding and keep their own supply of bull semen, which means that the number of career inseminators is dwindling. Some farmers have gone back to using bulls. Even if I had just four cows, I would buy a bull to service them in September or October and then sell the bull, repeating that practice every year and thus avoid keeping a bull on the farm year-round. It costs money to feed a bull all year, and often bulls become cross and dangerous.

Another possibility is to buy a young calf (or several), and raise it on artificial milk and pasture in lieu of mother's milk. Milk substitute is expensive and pasture calves do not grow as well on it in my experience. These calves, without mother's milk, will also need some grain to make good baby beef and the quality will not in any event be as high as that of calves raised with their mothers. But this is another way to grow your own beef and be sure it is not shot through with hormones and antibiotics.

If you begin to think seriously about a baby beef business on your farm, understand that you can get a good price for your meat only if you develop your own market. The commercial market does not recognize baby beef as a meat equal in quality to either the Prime or Choice categories that the beef industry and the government, hand in glove, uphold. So according to USDA grading standards, baby beef will not meet Prime and Choice price levels even though to many of us, it has higher quality. Because the whole traditional beef business is built on corn-fed fat cattle, a widespread baby beef business would be considered a distinct threat to the status quo. Why? The agricultural mainstream across the cornbelt is based solidly on corn and soybean production, as feed for cattle and hogs. The bulk of the farm equipment business and agrichemical business rises and falls with corn and soybean production. Moreover, the cattle are fed to market weight mainly in huge feedlots or hog factories controlled by influential people. The feedlots are most often situated far from where the corn and the calves are produced so there is much transport business in moving these supplies around. This monolithic meat establishment aims to keep things the way they are. An upstart who champions an agriculture where corn and soybeans and farm equipment and farm chemicals and truck transportation and huge feedlots are far less necessary, would hardly be welcomed. It is a shame because the production of 1200-pound fat steers is ecologically wasteful and financially inefficient. A baby beef business based on grass and local markets could generate a better rural economy, but with the government blithely stamping "USDA Choice" on the inefficient mainstream market product, it will be almost impossible for baby beef to compete until a million contrary farmers rise up to meet the challenge.

A cow in milk ought to get at least six to twelve pounds of grain a day in winter, but if she is eating plenty of very high quality hay or pasture, no grain is actually necessary. The husbandman uses his common sense and balances the ration between forage and grain depending on quality and availability. My own rule of thumb is to feed all the good hay a cow will eat. If the hay is not the best, then I add grain to the ration at the rate of six to twelve pounds per day, depending on the size of the animal and how much milk she is giving. Figure roughly that a mature cow, depending on size, will eat two to three tons of hay over winter in the north, less if stockpiled winter pasture or silage is available.

I do not wish to leave the impression that a small commercial dairy is not a good idea for a cottage farm. It is. One of my neighbors milks a dozen cows on only two acres of land and a little rented pasture in summer. He buys all his feed and the time saved allows him to pursue another career. This is the least-cost way into dairying since it does not require vast outlays of money for land and field equipment.

A new idea in dairying would also suit the cottage farm. Some dairymen using intensive rotational grazing (see chapter 6) are giving their cows prostaglandin injections that bring the whole herd into heat at more or less the same time, so the cows can be bred to have their calves at the same time—in spring, on pasture. The cows are fed mainly or sometimes entirely on pasture, and dried up in the late winter. This gives the dairy family a two-month breather from daily milking, at least theoretically. (There is always a cow or two who won't cooperate.) I refer you to the magazine *The Stockman Grass Farmer* (5135 Galapie Drive, Suite 300-C, Jackson, Mississippi 39206) as the best source of information.

The Homestead Hog

Commercial hogs are rapidly becoming the counterparts of the factory chicken: they are raised in huge confinement houses on contract with large agribusiness companies. Eventually, like the poultry growers, the hog producers will realize they are only factory workers, although not getting paid as well as other factory workers. Poultry growers are now organizing for a better deal from the companies to which they have become practically indentured and so eventually will the factory hog producers. The big companies are reacting very sharply to this new move because they know that when the grower starts receiving an income equal to what other industrial workers make, the price of factory meat would have to rise. That would allow small, independent farmers to compete with agrifactories even better than they can now. Last year an executive of giant Tyson Foods called the National Farmers Union, which has been in existence for years, a "bunch of socialists" for helping poultry farmers organize to seek a better price. Later he was forced to make a public apology.

The two pigs I am fattening now I bought as feeder pigs for $32 each. At butchering time, I will have in them about twenty-five bushels of corn and six bushels of oats, which cost me nearly nothing except my labor (see chapter 10). At two hundred pounds they will be worth at current market prices about $80 each. Even if I count all costs, I have a profit, not counting my labor, of about $40 a head. No Tyson Foods giant can equal that. Furthermore, we will butcher the pigs ourselves and get about two hundred pounds of meat—at a price of (conservatively) $1.25 a pound if purchased at the grocery—so our real profit, not counting our labor which is part of our way of life, is more like $100 per pig.

A cottager can raise hogs the traditional way, that is by keeping a few sows and a boar and raising two litters per sow per year. But I wonder if it is a good idea. A confinement operation costs too much and runs into odor and pollution problems. A pasture operation is much better, but it requires nose-ringing the hogs to keep them from rooting plus a well-maintained fence. I don't like the cruelty of ringing hogs, but unringed hogs can really tear up the terrain of meadow or woods. They manage to do quite a bit of rooting even when ringed.

A way to forgo ringing pigs and still keep them outdoors is to raise them in a field that you intend to plow up anyway. After the pigs get through with it, you may not have to plow at all. Just disk and level it. This is a particularly effective way to control weeds like bindweed and Johnsongrass. The pigs root up and eat most of the rhizomes.

We prefer to raise *a few* hogs in comfortable but cheap huts. If in the fall I have surplus corn that I know the other animals won't need, I buy enough weaned pigs weighing about forty pounds (feeder pigs), already castrated and given shots for various diseases, to eat up that surplus. The pen I built for the pigs is raised off the ground about two feet and the front two-thirds has a slatted floor so that much of the manure will drop through. The other third is roofed and kept bedded with straw. My pen will conveniently hold only three 200-pound hogs, but of course, if I want to raise more hogs, I can build a larger pen and even run two batches a year through it instead of one. I feed and water the pigs by hand, and clean out the front part of the pen every three or four days. Hogs help in this regard because they are naturally housebroken. They will not manure in their sleeping quarters, but will use the far end of the slatted floor, making cleaning simple.

A few pigs are easy enough to take care of this way. If I want to haul them to market, I can load them directly onto the pickup truck from their raised pens. That is the main reason the pens are raised so high— trying to run hogs or any farm animal up a slanted ramp into a truck is mayhem.

Keeping a boar for only a couple sows is hardly worthwhile. A way to avoid that, if you have a pickup truck and racks for it, is to buy a bred sow and keep her in a pen like the one I use for my hogs. The sow will have her pigs, usually eight to twelve, in the pen. Depending on the size of the pen, you can sell as many of the litter as necessary as feeder pigs when they reach thirty to forty pounds at weaning age, and sell the sow too at this time. Then feed out the remaining pigs to a market weight of 220 pounds and sell them. The bred sow will cost about $120. If five pigs are sold as feeders at $35 each, plus $75 for the sow, that's $250. The other five, fed out and sold as fat hogs, should bring a total of more than $400 at today's price, which is a little below average. That in round numbers amounts to something over $600 for a six months period. Buying two bred sows twice a year represents an income of $2400. But as usual, my advice is to try this on only a few sows. A larger number will mean spending more money on facilities and manure handling. Only as a very small, low-tech enterprise does this method return a large per-sow income.

Although I've never tried this myself, modern technology in hog production makes possible another slick alternative that the cottager can use to avoid both keeping a boar and constantly buying sows. Artificial insemination is somewhat easier for sows than cows. A sow's breeding period extends over four days. Boar semen can't be frozen like bull semen, but thanks to the wonders of UPS, you can order it as needed and get it the next day or sometimes the same day, along with an injection tube and other necessary paraphernalia for breeding. You should breed the sow on the second day of her period, and again on the fourth day. Unlike a cow, a sow in heat will usually stand still, especially if you straddle her. The inseminating tube is gently corkscrewed into her until resistance is met. I'm told that no special training is required, unlike the case of inseminating cows, but of course you will have to watch someone do it a few times to learn how. A cottager with a small, low-cost corn production program could keep as many as four sows year round this way and produce sixty or more pigs for spare-time income. Because artificial insemination allows you to use the best boars available, you could sell at least some of your hogs as breeding stock for higher than average prices. A good half of the feed could come from friends who save their table scraps for you, as ours do for us, instead of dumping them in the garbage disposal. Don't get greedy, though, and try to keep very many sows using this labor-intensive, primitive-facility system because the work of handling the feed and especially the manure will become overwhelming.

Draft Animals

The more a farm can produce its own power by biological means, the more it can claim true sustainability. The farmers of my father's generation all began farming with horses only, and to a man, the ones I've asked say that when the horses started leaving the farm, so did the profits. All those huge, magnificent brick homes that dot the landscape here, many of them now abandoned, were built with horse-farming money. "I'm spending up the money I saved farming with horses on tractors," neighbor Raymond once complained disgustedly at me from the seat of his tractor. And neighbor Junior Frey, who still keeps a team on his farm, told me a couple of years ago that, "if I just hadn't gotten started into tractors and expansion, I could have done just as well farming a small place with horses, but it is too late to back up now." Grandpaw Logsdon in his wheelchair in 1959 pulled me aside from a family conversation in which I was snickered at for saying I wanted my own land even if I had to farm with horses. He spoke low to me, so as not to be overheard. "Gene, I swear to God that if I were a young man now, I could farm a hundred acres with horses and come out farther ahead than these guys who are going whole hog into big-time tractor and bank-note farming." He was right, because most of those who dreamed of big time tractor farming in partnership with the banks ended up not farming at all. Some of them quit before bankruptcy and some quit after bankruptcy, and some ended up killing themselves. The number of suicides in Iowa in 1987 was 398, the highest number since the Great Depression. The suicide rate among farmers in 1983 was 46 per 100,000, approximately double the national rate for adult men. But the number of suicides in the mid-1980s was likely higher than what was officially recorded, as Osha Grey Davidson points out in his masterfully biting book, *Broken Heartland*. Many probable farmer suicides were recorded as hunting accidents and heart attacks because, you understand, good white, Protestant, Anglo-Saxon farmers do not kill themselves.

In the 1950s I was too young and stupid to see what was happening to us, and to see that we who remembered horse plowing were going to be the Last Farmers just as surely as Uncas became the Last of the Mohicans. It was only later that I started calling myself Uncas. But I did not follow my heart and get a hundred acres and a team of horses because I had neither the courage to run against the march of history nor the money to afford even that. But maybe you can. It is not a stupid idea, if you like horses. You would be in very good company. I could mention a score of Amish farmers who would make good role models for you; contrary to popular thought, they don't work any harder than tractor farmers and enjoy lots of recreation and travel if that's what you want. Maury Telleen is not Amish but he operates an exemplary cottage farm in Iowa with horses (and a little tractor) and has time to publish the *Draft Horse Journal*. Elmo Reed, during the years when he was a professional musician and professor of music at the University of Tennessee, also farmed with horses and today manufactures new horse-drawn machinery, especially a forecart equipped with hydraulics so that horse farmers can now use any three-point hitch implements made for small tractors. Down the road a few miles from me, Jack Siemon raises and sells mammoth jacks and jennets for the draft mule trade and makes good money at it. People come from all over the world to buy his mammoth jacks. Wendell Berry, one of our leading poets and essayists, farms with horses in Kentucky. Until he grew too old, neighbor Dale Kin raised and sold donkeys. On the other side of the county, Glen Keiffer has trained and worked oxen for many years. I worked one day in the field with him as he used his team to pull a corn binder. I much preferred their slow and ponderous movement to fidgety horses. Llamas are raised more for pets in this country or for guarding sheep, but in South America they are draft and pack animals. I have already mentioned that one llama raiser in Montana recently told me he had sold a female for $23,000.

And of course nearly every community supports or could support at least one farm that raises or boards saddle horses.

New, Rare, and Minor Breeds of Livestock

Interest continues to increase in offbeat breeds of farm animals, of which the pigmy pigs from the Orient are a good example. Some breeds of animals that have all but become extinct are now being patiently brought back into circulation. New crosses of minor and major breeds are becoming stylish as shifts in methods or markets take place. For example, Jersey cows are being sought again to give Holstein milk more protein. Dutch Belted and Milking Shorthorn are being cross-bred with commoner breeds of cows to produce an animal that can graze more efficiently—to take advantage of the widespread shift to intensive, rotational grazing. Beefalo supposedly combine the best of beef breeds and buffaloes. Other rare and even endangered breeds offer an opportunity for cottage farms. Joy Wind Farm Rare Breeds Conservancy (General Delivery, Marmora, Ontario, Canada K0K 2M0) is an excellent example of what can be done and a good place to write for information. The Institute for Agricultural Biodiversity (RR3, Box 309, Decorah, Iowa 52101) is another good source of information, as is American Minor Breeds Conservancy (Box 477, Pittsboro, North Carolina 27312).

Producing for the Pet Market

As urban populations move farther and farther away from indigenous nature, the demand for pets grows. Raising dogs, cats, and pet birds are enterprises well suited for cottage farms. The following is an example of what can be done, from a brief news item in a farm magazine of a few years ago:

A cottage farm couple, near a fairly big city, were selling baby chicks from their own incubators for 60¢ each. Also bantams from $5 to $10 each. Also parrots, parakeets, cockatiels, canaries, French poodles, German shepherds, and ducks. An Amazon parrot might bring $300; parakeets, $6. If human-raised instead of letting mother do it, a cockatiel might sell for $100, otherwise $70. Poodles, if small enough, went for $100; German shepherds from $25 to $100.

This farm also reared and sold a few beef steers a year, lots of baby pigs from their own sows, and live turkeys. Butchered turkeys brought more money, but that meant harder work. The couple neither advertised nor delivered. A sign in front of the place listed what was for sale currently and people stopped in, made their purchases, and hauled them away.

And now the punch line: *All this on four acres.*

Fish and Bees and Earthworms

Since even on a very small farm such as ours, there are more opportunities than any one family can pursue, I place great stock in fish and bees. Neither requires much attention. Both can provide food and fascination without the regular labor and continual feeding that domesticated animals require. They therefore make ideal "livestock" for cottage farmers short on space. If not densely crowded, fish in a pond (see also chapter 5) find enough food from their immediate habitat to forgo supplemental food. Society allows the cottager to harvest honey from all the land around, in return for the pollination the bees perform. I could add pigeons to this list of do-it-yourself homestead animals, since they too can scavenge most of their food on their own, as city rooftop pigeon raisers long ago realized.

If you eat fish once a week, fifty two-pounders or a hundred one-pounders should keep you supplied for a year, and that many fish you can raise in the backyard with a suitable tank and an aerator. A good source of information is the Alternative Aquaculture Association (P.O. Box 109, Breinigsville, Pennsylvania 18031). If you find fish farming intriguing, you can go from there as time and good sense dictate to selling fish to gourmets and fine restaurants. Since fish from "wild" waters are more and more contaminated, and since commercial fish farms seem bent on crowding fish to the point where more and more chemicals are needed to keep them healthy, the cottage farm could become the main source of fish clean enough to eat in the future. Dave Smith of Freshwater Farms of Ohio has been a pioneer in developing fish production tank systems in modified barns and cage systems in natural open waters and is an excellent source of information and supplies (Freshwater Farms of Ohio, Inc., 2624 North Route 68, Urbana, Ohio 43078).

The lazy beekeeper's way of caring for his bees probably makes the professional cringe, but it works for my purposes. I spend two days a year working with the bees, about a day repairing honey frames, and one evening getting the honey out of the combs and into jars. I leave plenty of honey for the bees to live on over winter and take only a small portion for our use. In other words, I let my bees do exactly what they do in the wild, only in man-made hives. Almost every year I read in magazines of some new viral, bacterial, or insect threat to bees, but so far (twenty years), so good for me. I once lost a hive to waxworms. I cleaned the hive thoroughly, captured a wild swarm, and started over. I keep two hives, because if one gets in trouble, or has a down year after the queen leaves with a swarm, the other hive is there to keep on producing.

Honey and honeybees fascinate me, not only because of the wonderful intelligence of these insects and the wonderful flavor of honey, but because I can procure their services almost totally free (not counting the cost of hives and equipment). Bees swarm every spring, and humans panic at the sight and call for deliverance. If the swarm lands on a low branch, it is fairly easy to capture. I set an open hive box under the branch, gently cut it loose from the tree, and shake the bees gently into the box. They don't much like being shaken that way, and buzz their disapproval, but they are full of honey and in too good a mood to attack, usually. Last year a few bees in a swarm did come after me, so now I wear my bee outfit as a precaution. As soon as the queen, buried down in the mass of bees, goes in the box, the rest will follow. After dark, I put a lid on the box, tape all openings and transport it to what passes for my bee yard. Sometimes I have shaken bees into a big cardboard box and then transported the box to the hive.

Earthworms are the livestock that all husbandmen keep as a result of and a sign of good farming. They can also be marketed at a profit, although I hesitate to stress the point because so many hustlers have taken advantage of this fact to cheat people.

Even so, earthworms may be just now coming into their own. Because of the ban on yard waste in landfills, more people are going to find it necessary or at least economical to compost their own leaves and grass clippings. Earthworms can help the composting process enormously, and in the process increase and multiply so that you can sell the surplus to other composters or to fishermen, or feed them to your backyard fish in place of that expensive fish food.

I offer all these ideas (there are SO many more) as examples of the imaginative ways you can add a little income and a little fascination to your cottage farm. Ignore university economists who tell young people that it takes a million dollars to get started in farming today. Seek out the people with horse sense and earthworm imagination and pay attention to what they say.

CHAPTER 5

Water Power

Rain makes grain.
Common saying among brokers on the Chicago Board of Trade

Ask an agronomist what plant nutrient is the most important, and you will be treated to a short course on nitrogen, phosphorous, potash, and a host of trace minerals necessary for plant growth. Ask a farmer that question and he will unhesitatingly answer: *water.*

Without adequate water, the most advanced agronomic blend of fertilizers or organic composts is so much powder in the wind. With ample water, there will be at least a fair crop even without any fertilizer. Add ample rain to fertility and your crop runneth over.

Ample rain means 35 to 40 inches per year. The real reason the United States has the most successful farming in the world is not because we have a so-called capitalistic economy and so-called innovative farmers, but because we have the greatest expanse of rich soil combined with the most advantageous weather in the world. Even the state-run farms in Russia would have done reasonably well in the Cornbelt. Already in the 1700s in the "Illinois Country," the French settlers around their frontier forts of Kaskaskia, Illinois and Vincennes, Indiana were raising surpluses of grain and competing with each other so sharply for the Spanish market down the Mississippi that they drove the price down too low for a decent profit. And this in spite of the fact that they farmed in a halfhearted, haphazard manner. They much preferred hunting, drinking, and chasing Shawnee women through the underbrush—probably the only traditional farmers in America ever to relax and enjoy life. Even way before that, in the Mississippian era of the moundbuilders, this land was so rich in food that Native Americans were able for quite some time to maintain stable, settled towns of considerable size without a domesticated agriculture. Within walking distances of their villages, they hunted and gathered all the food they needed.

I am forever surprised when I hear prominent individuals within the contrary farmer ranks say that they would prefer to farm in New England or California/Oregon if given their druthers. None ever say the Cornbelt. That just shows how contrary they are. There is a presumption that the Cornbelt is the domain of large scale agribusiness and therefore alien to a community of cottage farmers. There is also the coastal prejudice that nothing exciting or imaginative ever happens in the midwest. But what tremendous potential eludes the serious farmer or forester who decides to shun the richest soil and most beneficial climate in the world for food and fiber production. I am always deeply impressed by the awesome problems that good farmers of the dry West, the cold Northeast, the hot South, and the steep mountains overcome to achieve success. But if you are beginning as a cottage farmer and have no roots elsewhere, remember that compared to these places, farming in the Cornbelt is a little bit of heaven. Don't let the corn and soybean oligarchy take it all.

Regions with less than thirty inches of rain per year have to conserve moisture by leaving the land lay fallow for a year without crops, by mulching, by irrigation, and by growing crops acclimated to drier climates. The accumulation of salts in soil that is caused by irrigation has already ruined vast areas of former paradises in New Mexico, Arizona, and California. Farmers in regions of short growing seasons must spend most of their time storing food for winter as squirrels do, and shoveling snow.

Of course, if it rained only thirty inches and all of it came in the growing season, that would be ample for many crops. Rain that falls on frozen ground or on already saturated soil, conditions that often prevail in the midwest in winter, does not do us much good. Most of that kind of rain ends up in the Great Lakes or the Gulf of Mexico. As Virgil said, "Oh farmers, pray that your summers be wet"—but not too wet—"and your winters clear"—but not too clear.

Too much water is just as bad as too little, and I do not mean only flooding. In fact most flooding, here today and gone tomorrow, is not as disadvantageous to farming as soils that "lay wet" all the time. No matter how fertile they are, potentially, or how much fertilizer might be added, such soils will not produce crops until they are drained properly. If they do dry out by late May, they are planted late and so produce less and mature late or not at all.

If most of a field is dry enough at proper seeding time, farmers often give in to the temptation to try to cultivate and plant the whole field, wet part with the dry—especially when trying to farm more land than they can handle. Soil cultivated when it is too wet will not support a good stand of any of the grains or legumes, or good pasture grasses. It just produces gummy balls and slabs of soil that bake hard as rocks in the sun and grow up in milkweed and artichoke.

And the wet areas will grow in size. The reason? If wet soil is worked too wet, it compacts. The compaction causes the wet hole to dry even more slowly and the next year at planting time, the area of wetness will therefore be larger than the previous year. If you go ahead and try to work up a seedbed anyway, you compact yet a larger area, and so on, until what was just a little wet hole becomes a quarter acre.

Therefore many clay soils, however fertile, have to be drained, formerly with clay or cement tile, but now mostly with PVC pipe. Nonfarmers are often greatly surprised to learn that hundreds of thousands of miles of tile underlie the Cornbelt region. Even somewhat hilly land will have low spots here and there that require drainage.

One of the major tasks I face in bringing back my farm to good fertility is renewing the old tile lines that over the years of previous tenant farming were allowed to fall into disrepair. In fact tiling is more critical for me than for the larger farms around me. To lose an acre or two out of five hundred to poor drainage is not so bad, but even a sixteenth of an acre, out of a field that is only two acres in size, is considerable.

Tiling is almost miraculous in its benefits. In promptly removing excess water from the soil, the tile, placed two to three feet underground on a grade so that the water drains through the system by gravity, makes the soil more porous. Oxygen seeps through it more readily, and nutrients, especially nitrogen, become more easily available to the plants. Roots grow deeper and earthworms, which do not like sodden soil either, become plentiful. Soil acidity is more easily controlled by liming because the lime does its work of raising soil pH better on well-drained soil than on wettish soils. There is less temptation to cultivate when the soil is too wet, so compaction is avoided.

Some soils are so jackwaxy, as we say, that they can't even be drained with tile. Soils with such heavy clay usually occur on rather flat land of former marshes or lake beds where layer upon layer of vegetative matter solidified with the layers of clay particles settling out of the water. These soils require sophisticated surface drainage systems if they are to be cultivated. The Soil Conservation Service (the SCS, with offices in every county) can tell you all about such techniques (all about tiling too) but you would be better advised to plant this kind of land in grass and graze livestock on it or to build ponds and raise fish and ducks. South of my farm a few miles is a very large expanse of such clay, so tight that water won't go down through it to the tile. Or when it does, the fine wet clay seals the tile perforations. Humans very wisely have relegated most of this land to a State Wildlife Area.

All of which is a purposefully long-winded way of saying that the cottage farmer is a waterboy in every sense of the word. Water is as necessary to farmland as it is to a football team. A farmer helps nature bring the water in and when nature goes overboard (no pun denied) he must try to move the excess away as quickly as possible.

In choosing a farm, water should be the first concern. If nature doesn't provide enough, where are you going to get more? Build ponds, dig wells, practice desert farming. If nature provides too much, how are you going to get rid of the excess? Use tile drainage, which means making sure you have access to an "outlet"—an existing tile, creek, ditch, or pond low enough in your landscape to drain your tile system into. Or convert wet holes into ponds or wetland swamps and farm around them.

In choosing a garden spot, the same priority holds. If you garden with raised beds, you can, to some extent, avoid the problem of wet soil. Raised beds will work fairly well for vegetables even if you put them on concrete—Benjamin Albaugh was the first to describe how to do that in his 1915 book, *Home Gardening*. However, you commit yourself to a great deal of watering because raised beds usually won't stay moist enough through capillary action alone.

The spring-fed creek on our place was a major reason that we bought this piece of land. (The other reason was the woodlot.) Not only did the creek mean water for the livestock, but it guaranteed an outlet for tile drains. Two waterways (channels where water runs like a temporary creek after heavy rains) crossed our property and both had tile mains already in them to provide other outlets if needed. My first major chore was to return these waterways to grass so that they would not continue to erode into gullies.

As soon as money became available, I started putting in new tile lines in the bottom fields, with outlets into the creek. I hired a custom drainage engineer to do the work. I am now putting short tile lines through the garden areas, doing the digging by hand. Spading out a trench about two feet deep and a foot wide goes surprisingly fast—a good job for March when there is not much else to do. The slope of the land where I'm digging is pronounced enough so that it is easy to maintain a grade. My grandfather would keep a bucket of water beside him as he dug along, and every so often, would splash a little water into the trench and watch which way it flowed, just to make sure he was maintaining a grade. Two to three inches of fall per hundred feet is desirable though you can get by with less. The more grade of course, the faster the water will drain away.

With the ditch dug, I uncoil a roll of 4-inch plastic drain pipe and lay it carefully in the bottom of the trench, shoveling a little soil over it to hold it in place. After I make sure water is moving okay through the pipe, I cover it completely. If the tile goes through a particularly wet place, like at the corner of our garden, I will fill in with a layer of gravel over the tile and up to just below plow depth so that the water drains as fast as possible out of that area—what we call a French drain.

A drainage contractor with his ditcher digs and lays the PVC pipe all in one operation, guided by a laser beam to stay on grade. Then he pushes the tile ditch closed with a bulldozer. I'm doing the garden area by hand because the machines are too large to maneuver in the area.

In the middle section of the farm, drainage is also needed when I have the money. But in this case rather than outlet into the tile that exists under the waterway, which needs to be replaced, I have built a small pond and will outlet the new tile lines into it. At least that's my excuse for building the pond. I crave a pond like some people crave a seaside resort.

Enjoying Ponds and Creeks

Without a pond, a farm is sort of like a village without a church. A pond becomes the biological epicenter of a farm, drawing to it wild and domestic life looking for rest, relief, surcease, thirst slaked, a swim, a skating party, or maybe just a place to sit idly by and watch how nature pays homage to water.

Most of my sitting idly by has been done at the pond my father built fifty years ago just two miles away. A creek or river may be more fun to observe than the pond because the water moves, but either way, the water attracts not only the wildlife I encounter in the surrounding fields and the woods, but a different array of both plants and animals.

In the tree kingdom, the sycamore likes to grow close to water. Its wood makes the best butcher blocks—just saw off a section of log and put it on legs.

Black and peach-leaf willow, rarely seen in upland woodlots, shade the water's edge. You can often start willows, especially weeping willows, by sticking a green stem the diameter of a pencil into wet soil. Most of the time it will root. Willow makes terrible firewood, but artificial limbs are manufactured from it. Someone has to supply that market. Why not a cottage farmer? .

Weeping willows love to grow with their feet in the water. Along the creek, I have a huge one that is the first tree to leaf out in spring and the last to lose its leaves in the fall. There is always a day in late April when I can straddle a low branch, my back against the giant trunk, and in the space of an hour see a different species of wildlife every other minute: bird or bug or turtle or fish or muskrat or land animal. The willow's blooms in April are rather nondescript, but what a hive of insects they draw, the tree becoming one huge, throbbing buzz. The buzz attracts swarms of birds. A yellow warbler, fresh back from Mexico, hops among the branches, hardly pausing in its singing to snap up bugs. There is no yellow as yellow at that of the yellow warbler. I watch this one carefully, longing to find its nest. Normally this species goes farther north to nest, but you can never tell. I want to find a yellow warbler nesting because this bird has a reputation for knowing how to handle cowbirds. Cowbirds like to lay eggs in yellow warbler nests, among others, but the warbler is just as likely to build a new nest on top of the one desecrated by cowbirds, and then lay another clutch of eggs. Ornithologists have recorded instances of yellow warblers building as many as six nests, one on top of the other, to bury cowbird eggs. That I would like to see.

I always spot the red-winged blackbird in my big willow. A northern oriole always builds its hammock nest in a branch over the water. I look down now toward the water and am startled to see a redstart, black and flaming orange-red, back from Ecuador or maybe Cuba where it is called Candelita—"Little Flame." It is hopping about in a red ozier dogwood bush next to the water but soon moves on. Redstarts never sit still.

Then I see two snapping turtles glide through the water under the willow branches looking for crayfish which have all zipped, tail first, under rocks and logs to hide. The turtles slide under the pads of yellow water lilies that survive here against all odds. Schools of minnows drift by, and a kingfisher, clacking its raucous cry as it wings over the water, sends them scattering for underwater cover too. The kingfisher will not sit in my willow because there is no solitary dead branch to perch on and then plummet from, unimpeded by branches, into the water to catch a fish. Kingfisher goes on downstream to a dead elm stub over the water.

With my binoculars focused on the creek bottom, I see a freshwater mussel inching along on its one fleshy white foot, leaving a curvy line in the mud behind. If I stare at the mussel, it seems not to be moving. But when I look away and then back again a few minutes later, it has traveled a couple of inches beyond a little rock I was measuring its progress by. Its trail seems to be writing out a very casual letter S. For spring?

I hold my breath as a great blue heron drops out of the sky, folds its wings and begins to wade upstream on its stalky legs to do a little spear fishing with its long beak. A muskrat, poor of eyesight by nature, scrambles up on the bank under me and licks its paws. Our pet ewe, Bounce, also stops under the tree and peers querulously up at me. Humans are so weird, she baaas, and then shrugs hopelessly and goes on her way.

89

What I had thought was a nub of mud above the water's surface proves upon scrutiny with the binoculars to be a bullfrog nose. The rest of its body is barely discernible below the surface. I look upstream and suck in my breath again. A long-legged bird such as I have never seen before stands at the water's edge, its tail bobbing up and down in the most ludicrous manner. I leaf through the bird book I pull from my hip pocket. Eventually I find it: a solitary sandpiper. I watch it bob, wondering if solitary sandpipers always come to our creek and only now am I discovering them, or whether the winds high in its migratory sky blew it off course.

Beyond the bobbing bird, also next to the water's edge, a wild iris is blooming all alone, almost as if someone had planted it there. A little later a doe comes mincingly from the brush to drink. She smells human danger but does not see me until I have to blink and then she glides back to where, I'm sure, a fawn is waiting. Two male bluebirds flash around a hole in a fencepost along the field in the other direction, fighting for the territory. A pair of mallards floats under me, letting the current move them. And then float by two wood ducks, surely the most beautiful of the birds. What need have I for safaris, for continent-hopping, for far-off wilderness? I have hiked the wilderness trails of the world without leaving my weeping willow.

I'm distracted by a shiny green bug on a branch. I don't know its identity. I have only a bird book and wildflower book with me. I try to memorize the bug's features and color patterns so that I can look it up when I get back to the house, but I give up. There is a limit to what one mind can hold at one time and I have long ago reached mine.

My other major reason for building a pond of my own is that I love fresh fish. Though the creek provides a few, it is not deep enough for an ample supply and there is no way to keep the fish from swimming away downstream to the river. It is possible to raise fish in cages in larger creeks, but they usually must be fed regularly and removed in winter. For a typical warmwater pond in Ohio (one not fed by cold springs), largemouth bass and channel catfish are good choices. The former will prey on its own young and not overpopulate when intensively fished. The latter generally will not reproduce in ponds. If you must stock blue gills, use only the hybrid sterile strains since regular bluegills overpopulate and the large number of fry will tend to stay too small for enjoyable fishing or efficient butchering. Grass carp (not at all like the notorious "Egyptian" carp that overpopulate our rivers) are now touted as a good pond fish, and have long been a staple in Chinese agriculture. But they eat vegetation and will not prosper if stocked at rates higher than the amount of water plants available.

If you have a stream-fed or spring-fed pond where the water stays cold enough to support trout, (an average 60 degrees Fahrenheit in summer, never above 70 degrees) that should be your fish of choice.

You can feed fish in ponds similarly to feeding hogs or chickens in the barn. Commercial feeds are on the market. The Chinese throw fresh grass in the water for grass carp—hence the name. In oriental countries (and experimentally in the Unites States, especially under the aegis of Rodale Press in Emmaus, Pennsylvania and the University of Southern Illinois at Carbondale years ago), chickens or hogs are fed in pens positioned over the fish ponds. The animals grow on grain and the fish on the undigested grain in the animal manure that drops into the water.

Many techniques for raising fish as a farm product can be learned from catfish farmers and the literature thereof. But like all commercial farming enterprises, commercial catfish farms are forced, in order to make a profit, to industrialize their operations. Fish are crowded into ponds and chemicals must often be used to keep them healthy in such dense populations. While health officials say there is nothing to worry about, as they are wont to do, the cost of this high production makes those methods, in my judgment, too expensive and time-consuming for the cottage farmer.

In fact the best procedure for the cottage farmer to follow is to limit the population of fish (mainly by fishing) and let the biotic life of the pond provide all the food needed by its residents. Aim to sell just a few fish, if any. Maybe just sell fishing rights to a few acquaintances if you need to make a little money.

I have another possibility for my pond. It lies on upland pasture about a hundred feet higher than my lower fields. I contemplate siphoning the pond water in drouth seasons, onto at least the nearest of the lower fields, not only for irrigation but the plant nutrients from fish manure and rotting water plants. At the same time I would harvest the fish.

When the rains came and filled the pond again, I would restock the pond. This may or may not work out in practice, but is the kind of planning that makes life on the cottage farm so interesting.

Speaking of irrigation, there is normally little need for it here in Ohio, although extra amounts of water can work wonders, as we learned in one of our wettest years ever, 1992). I am ecologically against pumping out groundwater to grow crops we already have a surplus of. But capturing surface water in a pond to be used for emergency irrigation in a drouthy summer certainly transgresses no ecological commandments, especially since the pond will be promptly replenished by fall rains.

There is a no-cost form of irrigation that all husbandmen practice in their role as waterboys of the farmsteads. When livestock drink from the creek and then urinate on the pastures, they are in a sense carrying water, not to mention fertilizer, to the fields. This is an important point. When we talk of sustainable farming, we have to remember that even in grassland farming, where all the manure is returned to the land and there is no erosion, the animal products produced by the grass are removed: the meat, the milk, the hides, the wool, the eggs, and so forth. But when animals drink from the creek and the water in them is processed into nutrient-rich urine and put on the land by the animals themselves, there is an addition of nutrient value that helps make up for the loss represented by the meat and milk. This addition, along with rain and the nitrogen that legumes transfer from the air to the soil, as well as the plant residue that the animals don't eat, represents true sustainability if not a net gain in fertility without human labor. If you have ever watched a flock of sheep in action you know the urine/water addition is quite significant—about two pints of liquid fertilizer per sheep per day, I reckon. And cows of course, are mobile fire hydrants.

Farm ponds are generally of two types. One is a hole in the ground, to describe it simply, filled with run-off water or occasionally spring water. The other employs a dam to back up run-off water moving down a waterway between hills. The first kind is simpler to build and generally more satisfactory from the standpoint of maintenance. A dammed pond must be designed so that it is capable of handling the overflow from heavy rains. Designed wrongly, the dam can be easily and quickly washed away by flooding waters. The dam also must be designed so it doesn't leak.

There are plenty of qualified dam builders around. Again, contact your local SCS office first unless you are as independent as my father, who spurned all offers of advice and built his own dam, by damn. (It leaked a little and almost washed away a couple of times.)

An in-ground pond has no dam. If the overflow pipe is too small to handle peak floods, the water just rolls on down the grassed waterway like it did before the pond was there, no sweat. You do have to figure out where you want to put the dirt from the excavation. We employed a small bulldozer and operator to dig out our hole-in-ground pond. As the bulldozer pushed the dirt up out of hole, another man operating a front-end loader transferred it on to a low mound to the southwest of the pond. The mound of dirt now acts as a windbreak against prevailing westerly breezes. Some pond builders heap this extra dirt for a grassed privacy barrier if the pond is open to public view so they can swim nude if they feel like it. Cottage farmers don't get enough subsidy payments to afford to go all the way to St. Bart's in the Caribbean like the agribusiness farmers do.

A pond in this climate should be eight feet deep in the deepest spot so that fish have a good chance to over-winter. The shoreline should slope off almost immediately to at least three feet deep to discourage growth of water weeds, although that means little children can more easily slide in and drown. You have to kind of use your head and build to your own purposes. Three generations of children have played around and fallen into my father's pond and nobody has drowned yet. My uncle, too lazy to scoop snow off the ice by hand, tried to drown a tractor in that pond once, but failed. The water was only four feet deep where he went through. Several of us have gone through the ice while playing hockey, but hockey sticks are wonderful for rescuing such idiots. Pond safety depends not so much on the depth of the water but upon how sensible children (and their parents) are. My father firmly believed that "damn fools are going to get hurt or killed no matter how hard you try to protect them." He liked to tell the story of when he was a "little shaver" and an older boy threw him in the river. "I didn't know how to swim but he told me to swim and by God, I did, too," he said.

Livestock should not be allowed to wade around in the pond. The watershed above the pond from which water drains should be in grass or woodland so that dirt does not wash in with the run-off water. If you want to understand the power of water, build a pond with a watershed of cornfields. I watched one pond so situated that filled up with silt in a decade. Of course, it made a wonderful garden after that.

Above all, a pond should not have more than seven to ten acres of watershed for every acre of pond, especially if it is a dammed pond. You might go higher than that with a properly designed (and expensive) overflow system, but the gist of the matter is that a pond doesn't need much land draining into it. I know a very nice small one that maintains its water level with little more than a barn roof for a watershed.

If you can keep that little watershed free of pesticides and heavy fertilizer applications, your pond can remain fairly clean of pollutants and your fish safe to eat. That is its advantage over a creek or river where you have no control over pollutants. Also you have little control over flooding in creeks and rivers. Larger rivers are fun for the nature lover, but a headache for farmers whose land abuts them. Even creeks like ours, which I can jump across in some places in fair weather, breaks its banks in flood time and spreads over the lowland as much as a quarter mile wide in places. No fencing that runs perpendicular to the flow of floodwater will last very long. For these spots, I use board fences that I make and remake from old, used lumber or I buy old, used board gates at farm sales and continually replace them.

I used to fret considerably (Carol says considerably is not nearly a strong enough word) over these floods, which occur nearly every winter and occasionally in summer. But in twenty years all I have lost from floods were two cuttings of hay, and they weren't really lost because I chopped the muddy remnants up with the rotary mower to serve as mulch on the field. The high water usually recedes in three days. A few times water has completely inundated corn at six inches of height, but most of the stalks recover and go on to produce a crop. I have seen flood waters creep up the stalks of ripe wheat almost to the heads, and I still harvested the crop ten days later. Each flood brings a layer of my neighbors' rich topsoil and no doubt a little of their expensive lime and fertilizer to my lower field, so in the long run I am repaid grandly for the inconvenience.

Floods provide drama to farming. People who believe life is dull out here ought to try rescuing a flock of sheep trapped in a fence corner by high water. At night. With bolts of lightning for illumination.

Wetland Marshes

Prairie potholes, oxbows, hill seeps, frog ponds, peat bogs, cattail swamps, and various other kinds of marshland can provide a farm with a fuller range of biotic life than even streams and man-made ponds do. One of the saddest sins of humanity against nature is the draining of the prairie potholes of the upper Great Plains which has been going on for a century now and probably will continue as the grain gods further drain (whoops) the wetlands law of its effectiveness.

But let me speak to what I know, on my own stomping grounds. Two hundred years ago, this part of north-central Ohio was pockmarked with bogs noted for their cranberries. Sandusky, the name of rivers, streets, towns, and cities in this area, means "water within pools" in Wyandot Indian language. The bogs must have been noted for plentiful wild game too, because around these old depressions now plowed for corn and soybeans year after year, I find many flint arrowheads—twenty-eight in an afternoon, once! I say "depressions" because that is all that remains today. Many of these depressions are so low in the landscape that they cannot be adequately drained to tile outlets, yet farmers stubbornly keep trying to farm them whenever they dry out enough to plow. The two that I have watched and studied closely for fifty years have rarely produced a profitable crop. Had they been preserved as wetlands, they would have produced more wild food from hunting, fishing, (even cranberries again?) than they have ever produced in surplus grain.

As each generation arrives with yet more powerful digging tools and money, new attempts are tried to get tile deep enough under the old bogs to drain them, and I suppose eventually someone will succeed. If so, then the dried-out, powdery-muck soil will blow away. As it is, even with all the technology thrown at these bogs, they lay under water or are totally soggy about half the time. How much saner it would have been to leave these potholes as marshland, or at least excavate them a little deeper and turn them into ponds. The ponds or marshes could have been used as tile outlets for surrounding fields at far less expense than running big eighteen- and twenty-inch tile mains twelve to fifteen feet deep for several miles to a creek to drain the old bogs. The irony of the situation is that these potholes are not protected by the new wetlands law, since they have a history of being "farmed," while the wetlands law does apply to piddly little mosquito breeding puddles that can be easily drained for decent crop production. Farmers with a "look-the-other-way" policy on the part of local officials have cleverly gotten around the law in most instances anyway. The ultimate result from a poorly written law is that in this county very few worthwhile wetlands are saved while much time and red tape (with who-knows-how-many under-the-table deals) are wasted to allow farmers to drain the wet spots that are insignificant as wetlands anyway.

The most ludicrous example occurred when the local landfill was forced, by the law, to build a "new" wetland because in the course of expansion, it had to destroy a woods with a bit of swamp in it. The spot picked for the "new" wetland was a field of prime farmland where a new and complete tile system had only recently been installed. The landfill owners had to spend thousands of dollars scooping out a shallow basin of several acres with a bulldozer, destroying the tile installation in the process, and piling the dirt into a steep hill. The hill buried several more acres of farmland and will itself be too steep for anything except a toboggan run and it will erode back into the "wetland" from now until eternity. The site of the "wetland" would have been much better used for a woodlot of valuable black walnut trees which this land was especially well suited for. And if a swamp were indeed desirable, like the one the landfill had destroyed, all that was necessary in the first place was to plug the outlet of the tile system draining the field. That would have taken ten minutes. The country is losing its common sense because pompous asses in high places will not allow local people to solve local problems in a manner practical for local situations.

We are obviously faced here with another example of socially approved insanity, and the only effective solution I know is for more cottage farmers to buy more land. I've waited a lifetime to buy the field in which one of the potholes I've described lies. I have seen what a gram of this soil looks like under an electronic microscope—a world teeming with a splendorous explosion of technicolored microorganisms that reminded me of a tropical rainforest. I would like to turn that wetland back into a cranberry bog and duck pond and sell hunting rights, which would be a more profitable enterprise than mulishly trying to grow corn there. And I would still have the other half of the field to turn into productive cropland by draining it cheaply into the bog. My chances of acquiring this land are extremely slim, but I suggest the idea in case others are faced with a similar opportunity. Such land usually sells cheap because crops fail on it and houses would sink in it.

Then again there is sometimes reason for optimism. Prosperous farmers have donated one such old bog to our village. It lies right at the town limits, and will become a demonstration wetland for all the public to watch and learn from. I am even hoping for cranberries.

A list of the possible "crops" that freshwater wetlands can grow would fill three or four chapters. But here are a few:

Cattails. The roots, or rather the rhizomes growing from the roots are delicious. I lived on them for two days while doing survival camping along a Minnesota lake when I was young and—fortunately—foolish. Moreover, scientific studies done by Dr. Leland Marsh of the State University of New York years ago indicates that an acre of cattails will produce thirty tons of flour from the rhizomes and leave enough root in the ground to provide for the next year's crop. Cattail leaves are durable enough for woven basketry. Best of all, "cattail corn," the seedheads picked green and steamed like sweet corn, is a delicacy of fancy restaurants who pay $8 a pound and up for them.

Wild rice. Researchers have told me that there is no real reason why it wouldn't grow, at least for a modest family supply, in an Ohio wetland.

Watercress. Although it prefers to grow in running spring water, a very contrary farmer I know has made good sideline income growing watercress in a marshy pond fed by springs. This same farmer has capitalized on water power in what is perhaps the most basic way. He sells his good-tasting, unpolluted well water by the bottle out of a little roadside stand. "Many people no longer have any water to drink except that awful-tasting, chlorinated city water," he explained.

Crayfish. I doubt whether a crayfish farm can be really profitable in northern Ohio, but there are plenty of them in Louisiana. And our Ohio creeks, polluted as they are, still support crawdads. I remember as a boy seining enough of them to fill a three gallon bucket in two sweeps. Yes, you fix them just like lobster and they taste very similarly. Let's see.

At today's lobster prices, a three gallon bucket of crawdads would be worth . . .

Waterlilies. Many water gardeners have built small sideline businesses selling waterlilies that they raise in ponds or garden pools. The yellow waterlily has grown wild in our creek for at least sixty years and I presume they were there when the Mississippian Indians built the mound that rises nearby. Maybe it was the moundbuilders who planted the lilies in the first place.

Bullfrogs and snapping turtles. These two amphibians will proliferate in a wetlands pond if they are harvested in moderation. Both are excellent eating.

Muskrats and their predators, minks. The fur of muskrats was once prized in the fur trade and mink still is. Once wildlife lovers are educated to understand the dynamics of population growth, fur coats will not be so wrathfully condemned and then these furs may again become a sideline source of income for small farmers. In addition, muskrat meat is very tasty.

Ducks. Ducks of course can be marketed for the meat. Not fifty years ago, small farmers with access to ponds and rivers made ducks a major source of income. That was not all good, however, because often so many ducks were crammed into a small area that waterway pollution become a problem. "All things in moderation" . . . oh, if humans could only learn that.

Duckweed. This pond weed appears to most of us as an ugly green scum on pond water. To Viet Ngo, a Vietnamese turned Minnesotan, it was the basis of a business. Duckweed turns out to be an efficient purifier of wastewater and a protein feed for cattle exceeding even quality alfalfa in food value. Viet tells me it could be a good human food too, as it is in other parts of the world.

Peat bog products. Needless to say, true peat bogs or any marsh with a high acid content in its water and soil can contain plant life markedly different from land around it. Pitcher plants and the bug-eating sundews, for example, are easy to sell as houseplants, but many bog plants are endangered species so don't sell those if you don't know how to grow more. Also sphagnum moss, the old, partially decayed deposits of which become peat, is most useful in pot mixtures for container-grown nursery plants. Dry sphagnum moss has twice the absorbent capacity as cotton and would probably make an excellent diaper lining especially since it also has medicinal value. Herbalists say it makes a good temporary bandage. Just think, a throwaway diaper that really is environmentally-benign.

Milkweed. Swamp milkweed, the one with the dusty pink flowers, prefers wetter soils and often grows in wetlands areas. The pods of milkweed floss (the little parachutes attached to the seeds that allow the seeds to float so well on the wind) have definite commercial possibilities. The floss was used in World War Two as the packing in life preservers and is now used, with goose down, in comforters made by the Ogallala Down company of Ogallala, Nebraska. The company's standard ad says: "WE'RE GREAT IN BED." Herb Knudsen, Ogallala's president, told me in 1990 about research at the universities of Kansas and Nebraska indicating that revenue from growing milkweed is approximately the same as from corn. I imagine that is even truer today because in some years cash grain farmers net next to nothing from corn without the government subsidy (see chapter 8).

I but ripple the surface of the watery potential. What your pond will draw above all else in the eastern half of the nation will be Canada Geese. This is currently not good news, because these big birds are rapidly overpopulating and becoming a nuisance. Well-intentioned laws protect them too much, for one thing. Hopefully, society will eventually get past its overly protective attitude about wild animals and support "sustained yields" rather than the current deep denial of nature's awful but necessary means of survival: All creatures, great and small, must have predators. When they do not, humans must sometimes step in with a little management and control the population. Nature even dictates that humans have predators. When they do not, and refuse obstinately to use their rationality to control their own population, history shows that, century after miserable century, humans will prey on each other.

The real problem in wildlife management is one of too much government-from-on-high which, as in the case of wetlands, cannot take into account, when making general regulations, individual or local differences. Fundamental to that problem is the governmental attitude that local people aren't capable of making the right decisions for their local situations. We locals are deemed to be inherently loco, or ignorant, or worse, outlaws who must be forced to do the "right" thing, while the enforcers consider themselves infallible. Dreadful. Millions of us are out here on the land because we love it, and we understand that diversity is the key to ecological health. We intend to protect that diversity even to making sure there are enough damn groundhogs around. But we also know best when too much is too much. I have never killed a Canada goose and face the prospect with loathing, just as when I kill a hog. But I know better than a tableful of faraway bureaucrats when Canada geese can and should be killed to improve the environment of my local area. The government, instead of sending its Rambo game wardens to harass us, should be intent on giving us good general information to help us in our decision-making. But bureaucratic regimes can't even do that. The few scientists who have fought hard for the correct understanding of wildlife population dynamics have had to risk their careers to do so, and their documentation is still not heeded by the regulators. As a result, wildlife agents are forced to follow a hypocritical and stupid method of Canada goose control: they spend costly time breaking eggs in the nests. It is okay to break eggs, not okay to kill birds and give the meat to people starving from malnutrition.

The Water Power of Mulch

It seems strange to talk about mulching gardens in a discussion of water but mulching is one of the most effective forms of water conservation and irrigation. Mulch slows the evaporation of soil moisture. This can be a critical factor during temporary dry spells—as effective, or even more so, than irrigation. Mulch conserves moisture not only directly but indirectly by increasing organic matter over the years. USDA statistics say that a soil with 4 percent organic matter can hold six inches of rain before run-off occurs. (Most soils after years of hard farming contain 1 to 2 percent organic matter or less.) Where irrigation is a necessary adjunct to farming, installing drip hose with tiny holes in it for irrigation under mulch can mean tremendous savings in water and money. For raised-bed growers in water-short areas, the combination of mulch and drip irrigation is the only profitable and environmentally friendly method.

One of the most thoughtful large-scale farmers I know is Ted Winsberg in Florida. Ted grows about two hundred acres of sweet peppers on his farm. That is a lot of peppers. Lately he has been wondering about the conventional manner in which peppers are raised on a large-scale level—on beds covered with plastic mulch and treated with methyl bromide to kill weed seeds and nematodes. In this system a very sophisticated and expensive method of irrigation is used: a network of ditches which maintain water under artificially-shaped beds at just the right level. The beds and some of the channels have to be reshaped with earthmoving machinery every year at considerable expense.

So he switched a portion of the farm to permanent beds mulched with compost and watered by drip irrigation. No more expensive bed-shaping every year. So far, no more methyl bromide either. "It's kind of ironical," he told me last year. "I've been raising two hundred acres of peppers while my son out in California raises a half acre in a greenhouse. He clears more money than I do."

Springs

There are creekbed springs upstream that keep our creek running year-round, and I hold my breath that they do not dry up like all the hillside springs in this vicinity have done. A clear, clean spring can be the biggest asset of all to a cottage farm. When I think of springs I think of the Kuerner Farm at Chadds Ford, Pennsylvania, where Andrew Wyeth has painted most of his most famous temperas. The farm would be an amazing place even if Wyeth never painted it, which I presume is why he did paint it. Water for the house, springhouse, cisterns, and barn is piped from a hillside spring that is higher than the buildings. The water has been running through the farmstead for at least two hundred years by gravity. Anna Kuerner, still very much alive in her nineties, used the water to cool the milk and butter in the springhouse for many years, and Karl, her husband, now deceased, stored his apples and his famous cider there. Karl, Jr., now operates the farm with help from his son Karl, who is a masterful artist himself. I have the great good fortune to have known all three generations of this remarkable family. "Occasionally we have to clean off the screen on the pipe inlet," says Karl, "but other than that the system pretty well takes care of itself." Imagine a never-failing, never-ailing water supply and refrigerator. Have we really progressed? Washington and Lafayette planned Revolutionary War battles in that house and no doubt drank the same spring water.

Another spring, or possibly another upwelling of the same vein of water, provides the water source for the pond in front of the ancient Kuerner home. For many years this pond was the Kuerner kids' swimming pool. It is the centerpiece of many famous Wyeth paintings, especially "Brown Swiss."

"Sometimes we just jumped in the horse trough in the barn in the old days," says Louise (Kuerner) Edwards, who with her son still operates her own farm about twenty miles away. Wyeth has immortalized that horse trough, too, in his painting, "Spring Fed." The reason for Wyeth's popularity and influence is much argued by art historians, but seems simple enough to me. Human culture is still rooted in farming and Wyeth has chosen rural life as his medium, so to speak, for expressing the emotional upwellings that charge his creativity with energy. The millions upon millions of people in the world who are contrary farmers or the offspring of contrary farmers, pining for something they associate with the way their parents or grandparents lived, can simply enjoy Wyeth's works as very striking illustrations of that life, or else as the deeper outpourings of the universal human spirit within that life.

For example my favorite Wyeth painting is "The Virgin," a rendering of teenaged Siri Erickson standing naked in her father's barn. What makes it my favorite is the story that goes with it, told by Wyeth himself in *Two Worlds of Andrew Wyeth: Kuerners and Olsons*, a collection of his works published by the Metropolitan Museum of Art in 1976. As Wyeth was painting Siri in the barn, he says, she became intent on something she saw out the barn door. Suddenly she rushed out, stark naked, grabbed a club and sprang into the garden where a groundhog was eating vegetables. "She just clubbed it to death," Wyeth is quoted as saying. Then she came back and resumed her pose, groundhog blood splotched on her leg. All in a day's work. Wyeth said he was astonished but considered the incident "a bit of luck" for him as an artist.

But it wasn't luck. He *lives* on the land he paints every day, enabling him to see the real rural culture, out of which Siri sprang, and to paint it honestly. "She once told me she liked to ride bareback in the summer at nighttime completely nude with her blond hair streaming behind her," he also relates.

Andrew Wyeth, though certainly not a farmer in the strict sense of the word, is to me the most contrary farmer of us all. He understands that our life is not dull and boring, but full of drama. He understands why we will be here on the ramparts, clubbing groundhogs to death if we must, riding naked in the wind if we feel like it, long after the tinsel of civilization passes into hydrocarbon history.

CHAPTER 6

A Paradise of Meadows

To make a prairie it takes a clover and one bee,
One clover, and a bee,
And revery.
The revery alone will do,
If bees are few.

Emily Dickinson

I encountered an apparition in our pasture field on a recent November morning. A ghostly white cattle egret was strolling along with the sheep as they grazed. Egret? I stared unbelievingly, then ran for binoculars to verify the identity of the chicken-sized bird. American egrets are common in the South where farmers call them cow egrets but are as unusual as alligators in a northern Ohio sheep pasture. The bird book confirmed my calculated guess: egret it was. How it got here I have no idea though I understand this species is drifting more frequently northward than it used to. How long this one will stay in our meadow before it falls prey to cold weather or a coyote or a fox I can only guess. For the present it walks haughtily about like Theodore Roosevelt on the White House lawn, occasionally seeming to peer into the ewes' eyes with the intensity of an eye doctor in pursuit of a detached retina. Absorbed in its ministrations, it does not even mind when one of the ewes butts it away. When it runs out of anything else to do, it gobbles up chunks of manure, both sheep and cow, the way a chicken will sometimes do. Egrets are prized by southern husbandmen because flocks of them help control insects that buzz and bite and bother the animals. Stockmen sometimes build roosts for the egrets in their stock ponds because the birds like to perch on dead trees surrounded by water.

Our lone visitor will have to make do with a tree perch over the creek. And coming in November, it will find precious few bugs though there are always a few brownish-yellow dung flies on the cowpies this late in the year, and earthworms underneath.

I think of the egret as one of the many bonuses we have received for following what we believe to be the basic tenet of sustainable agriculture: The grazed meadow, or pasture, or grassland—call it what you will—is the ecological and economic foundation of farming.

We are convinced because of the advances being made in what is called controlled or managed *rotational grazing*, by which meadows or pastures are divided into sections or paddocks and grazing animals moved from one to another as necessary for the most efficient growth of both animals and pasturage. When controlled rotational grazing is done properly, the meadow or pasture can provide almost all the food for the animals on the farm, and some "grazing" for humans, too. Have you tried wilted dandelion bud salad with vinegar and chopped boiled eggs? Or broiled *Agaricus* mushrooms basted in butter and lemon juice? Egrets and who knows how many other manifestations of wild nature provide the dessert.

The reason controlled rotational grazing, or *grassland farming*, the older term which I prefer to use, fits the cottage farm so well is that this is becoming the most economical way to keep animals as the cost of annual crop farming rises. In grassland farming, the animals do most of the work of fertilization, weed control, and harvesting. Since the pastures are re-seeded only rarely, planting costs are minimal. Also, grassland farming provides hay as a by-product, and hay, if it must be purchased, is the single biggest expense of the small farm that keeps animals. In fact, on a very small farm such as ours, it is hardly possible to make a profit on animals if you must buy hay rather than make your own. Most important of all, grassland farming with the majority of the acreage in sod is the only "no-till" system to effectively control soil erosion, the ruination of civilizations. The highly touted (by agribusiness) chemical no-till method, where toxic poisons are used to denude the land of weeds so that crops can be planted in undisturbed soil, is seemingly less effective and in any case too new to have proven itself. It appears that as weeds continue to develop immunity to weedkillers and as harmful insects continue to proliferate in the undisturbed soil, requiring the use of lethal insecticides, chemical no-till will prove to be only a temporary solution to erosion.

To understand a meadow, you really need to sit down in one a while. Maybe like for twenty years. But instead of lapsing into reverie right away, pay attention to what's going on around you. You should be able to reach out and touch several different species of plants from your sitting position, and spy, in the space of a quarter acre or so, perhaps twenty or thirty different ones. Generally, the older the pasture (the more years it has not been plowed up and renovated), the more plant species can be seen there. It took fifteen years from the time I started my pasture, on land which the former owner had devoted to annual row crops, for blue-eyed grass to appear, the most delicate and to me beautiful of our native wildflowers. I have no idea how it came to grow in that field. Did a bird eat a seed somewhere else and drop it here in its manure? Perhaps we should leave the explanation to magic. Magic is more fun than science, and maybe in the long run more reliable.

As I sit in a meadow, savoring the magic, I can tell the season by the plants that are flowering. Dandelions and spring beauties dance across my meadows from late April to late May. Violets hide among them. (Our friends Dave and Pat have naturalized daffodils in their pasture.) White clover blooms profusely from late May through June and less vigorously the rest of the summer and fall. Wild strawberries bloom in early May. Cinquefoil, which looks somewhat like a strawberry plant but has a yellow bloom, comes in June. So do buttercups. Although their blossoms are hardly noticeable, many of the early grasses, including mainly bluegrass, bloom also about the beginning of June. The little blue (and sometimes whitish) blossoms of blue-eyed grass nod shyly in June. Red clover, alsike clover, hop clover, alfalfa, and other legumes all bloom, purple to pink to white to yellow, in June, if not grazed off by the animals. Grazing is good for clovers (or more correctly, good for the grazier) as grazed clovers regrow quickly. If allowed to go to seed, clover plants will not make a strong second growth. The weedier and more undesirable meadow plants start blooming in July and continue until fall; here in Ohio that means sourdock, Queen Anne's lace, thistles, mullein, horse nettle, burdock, sheep sorrel, milkweed, yarrow, plaintain, and many others. In the fall ironweed, goldenrod, joe pye weed, and wild asters bloom—not desirable for grazing either, although pleasant to the eye. Most if not all these plants, while a weedy nuisance if over-populated, are revered for one or another curative power in traditional herb medicine and so may benefit livestock in ways not precisely known yet. I am not sure that boiled root of ironweed is good for motherfits, whatever that is, or that yarrow will purge the kidneys, but neither am I about to scoff. Many of our medicines were discovered by way of herbal folklore. The animals know. When allowed to live in their habitat unbothered by humans, how many mountain goats, elk, deer, rabbits, buffalo, or groundhogs suffer from motherfits or need their kidneys purged?

A meadow is more than just the plants and their variously splendored blooms. The greater the variety of plants, the greater the variety of insects that can be seen humming above them. And the more diverse the insects, the more kinds of birds will come to catch them. One of my pasture delights is watching the barn swallows swoop low over me when I mow—the chugging tractor and clattering mower chase bugs into the air, and the swallows know. As soon as I enter the pasture with the tractor, they wing into view from nearby barns.

In my boyhood, I spent hours in a meadow a mile from the one I am now describing, watching the tumble bugs (dung beetles or scarabs) form marble-sized balls of manure from pasture droppings. Mother beetle would deposit an egg in each ball. Mother and father beetle would then roll the ball a few yards down a sheep path and bury it in the adjacent sod. Dung beetles can get rid of prodigious amounts of cattle manure, researchers say, and in doing so they help control pesky face flies, which also lay eggs in the manure. In Texas, the beetles have been introduced to cattle ranges where they effectively bury droppings that would otherwise smother out significant amounts of grass temporarily. Strangely the beetles have disappeared here and no one can tell me why for sure.

Other meadow insects I've watched since boyhood are still here, their food chain connections to both plants and birds intact. Grasshoppers and crickets still provide one of the main sources of food for birds such as meadowlarks, song sparrows, bobolinks, bluebirds, pheasants, and quails. Bumblebees and honey bees love the clovers with their abundant stores of nectar. The elegant kingbird tries to catch the bees but usually gets only the slow-moving drones. On milkweeds dine red milk-weed beetles and monarch butterflies. On dogbane, a milkweed family plant which is somewhat poisonous, the breathtakingly iridescent copper, blue, and green dogbane beetle chews away. The birds learn not to eat monarch butterflies or any bug that eats milkweed, since these insects taste of the bitterness of the milkweed juice. Or so naturalists reason.

In the meadow you will also see occasionally a viceroy butterfly darting about, and the viceroy keeps a secret that science cannot yet unravel. The viceroy's larval form is so ugly it suggests a bird dropping, and its chrysalis is not much more attractive, while the monarch larva is a crisp black-ringed green in color and its chrysalis glistens like a jade jewel. The viceroy larva feeds mainly on willow, aspen, or poplar tree leaves; the monarch larva feasts exclusively on milkweeds. The monarch migrates; the viceroy does not. Yet the two insects look almost exactly the same when they metamorphose into butterflies. The common explanation is that the viceroy evolved this way—by mimicry protected from birds, who mistake it for the bitter-tasting monarch. But so many conditions would have had to be present concurrently through the parallel evolutions of the two insects that this explanation hardly does more than beg the question. The truth is that no one knows why the viceroy and monarch are look-alikes. It's meadow magic, I say.

Queen Anne's lace or wild carrot draws the gorgeous Eastern black swallowtail butterfly. Eastern tiger swallowtails and spicebush swallowtails wander into the pasture sometimes, alighting on thistle and clover blooms. More often you see the painted lady butterfly on the purple blossoms of bull thistles and Canada thistles. The smaller sulfur and copper butterflies can be depended upon to reward the sharp eye if nothing else is in attendance.

The birds mentioned earlier and many others are on the lookout for not only the winged insects but also their larvae. The birds, especially quail, walk through the grass looking under weed leaves where worms like the potato beetle worms that chew on horse nettle, a relative of the potato, think they are well-hidden. Flickers patrol the meadow floor for ants.

High above wheel the hawks and buzzards. The buzzards hope that I will shoot the groundhog that has emerged from its hole on the brow of the hill. Buzzards find dead groundhog to be a tasty morsel, especially the rodent's eyes which the witchy vultures invariably pluck out first. Buzzard hors d'oeuvres. Or perhaps they circle above me because they think *I* am dying. They are not accustomed to humans who sit still in meadows. They wait for my breathing to stop, relishing my eyes too.

The hawks scream, protesting my presence. They want to come wheeling low and swiftly over the field after mice or slow-moving mourning doves, but not while I am watching. Their vision is so keen that even from far in the sky, I'm certain that they can see my eyes blink.

The addition of livestock to the meadow (replacing the wood bison and elk of earlier times although the deer have come back) also draws more wildlings. Cowbirds perch on top of cows and sheep, hankering after flies and ticks. Ground wasps whiz over the cows, grab flies, and take them to their underground nests along the pasture fence for larval food. Close to water, damsel flies and dragonflies hover above the grass, hunting mosquitoes that are lured by the cows and the promise of blood.

The most comical sight in the meadow is mother skunk, followed in single file by her three children, all of them grunting and complaining about problems known only to skunks. Body odor, perhaps. They tip over cowpies for the earthworms underneath. At night they dig up the ground wasp nests and eat the larvae. Woodcocks also tip over cowpies and plunge their stout beaks into the soil for nightcrawlers sure to have slithered down their holes the moment the cowpie moved.

Know that I am but playing half way through the overture to this wondrous piece of meadow music. I won't last long enough to play the whole score. But I continue to play at it, in reverie or at work, perhaps hoeing in the corn patch nearby if not nestled in the grass, my ears as well as eyes taking in the harmonies.

The ecology of this small, eight-acre pasture is a puzzle that can never be completely assembled. Every time I fit a piece into its proper place, new holes open up. As a puzzle, the meadow's interrelated forms of life remind me of the fractal clusters that Chaos scientists conjure up on their computers, seemingly random patterns that merge and emerge into beautiful, orderly designs on the computer screen. All my life I fill in random pieces while I wander dazedly through fractal patterns that are real, not computer-drawn. The missing connections multiply faster than the found ones. I do not care. It is here, in these empty spaces that magic reigns.

Managing a Meadow

Reverie will not put food in the baby's mouth. I observe the natural life of the meadow not only for pleasure, but to learn what I can about how to combine this wildness with domestic animals, to the advantage of both. Obviously, this kind of management is more passive that active. Essentially, I divide the pasturage into parts (paddocks) and then I and our grazing animals take turns mowing the sections. The "what" to do is simple; the "why" and "when"—the knowledge behind the "what"—entails great ecological and botanical complexity, never wholly grasped. If you want to get commercially intensive with controlled rotational grazing, a good source of information is *The Stockman Grass Farmer*, already mentioned in chapter 4. For the cottage farm, I believe in a more laid-back approach.

The reason I advocate moving forward slowly is not only because I do not believe that short-term profit should be the cottage farm's first goal, but because without a history of experience in this new kind of farming, we do not know its potential. That is why grassland farming is so fascinating and challenging. This approach changes farming from a machinery-based activity to a knowledge-based activity in which the graziers' sensitivity to the natural world is at least as important as the scientific facts that guide them. If agriculture is an art, and it is, then grassland farming is the most artful kind of agriculture.

The artful complexity is twofold. On the first level of management, you consider the mixture of grasses, weeds, and legumes in your pasture from the practical perspective of farm production. When are the plants ready to be grazed? How long can they be grazed without harming their vigor for regrowth? How many pounds of meat or pails of milk or dozens of eggs will an acre of grasslands produce? Which are warm season plants, those that grow well in hot, dry summer, and which are cool season plants, those that grow profusely in wetter, cooler spring and fall? Which grasses can be "stockpiled" in summer (that is, left ungrazed) for winter pasturage? Which grasses and legumes grow best in a particular climate or soil? Which can be counted on for a hay crop in addition to grazing? (The answer to all these questions is, "It depends," but I shall try to be more helpful as I go along.)

On a second level of consciousness, if you will, you must consider the relationship between grazing as an agricultural practice and grazing as a way to maintain ecological wholeness. In this regard, the first important "law" to keep in mind is that where rainfall is plentiful enough to support forests, *grazing makes the meadow*. Seedling trees quickly populate a meadow in woodland regions, and will in the space of about forty years turn the meadow into a forest unless the seedlings are killed by fairly heavy grazing and routine seasonal mowing. Sometimes, as in the case of white thorn (wild hawthorne), hand-weeding is necessary. White thorn is the characteristic seedling that, in this northern Ohio region anyway, comes first in the normal succession from grass to briars to forest, and not even grazing and mowing will stop it. I cut them down as grown trees, and cut out seedlings with the spade when they are tiny. White thorn makes a pretty tree actually, and in improved, horticultural varieties is a favorite ornamental. Seeing one on the lawn of a courthouse in a large city, Dave, my neighbor, stopped dead in his tracks and muttered: "Where's my axe?"

A second "law" of controlled rotational grazing is that, without rotations, even a few animals on an uncrowded pasture might eventually depress those plants they like best because they wander over the whole pasture in search of their favorites, like a child choosing strawberries over broccoli. By dividing the pasture into small sections or paddocks, the animals have to eat their "broccoli" as well as their "strawberries" before being moved to the next paddock dining table. And by the time they graze through a succession of paddocks and get back to the first one, the tastier plants as well as the less tasty will have had a chance to grow back again. In other words, by rotating the paddocks frequently, plant diversity is maintained. Diversity means not only more species of life in the pasture, but a more balanced diet for the grazing animals. Exactly when to move the animals so as to maintain diversity while encouraging the most efficient regrowth is a very fine art that resists precise instruction. I tend to move the animals sooner rather than later and then finish the "grazing" of the rougher, less tasty plants with the mower. I try to favor my blue-eyed grass by reserving the paddock where it appears for mid-summer pasturage, after the wildflower has gone to seed.

Among the other management questions to be addressed, these are the most important: Can I reseed the pasture without tearing it up and exposing it to erosion? Yes. Graze the sod to be renovated rather severely in the fall so that the ground is almost bare. Then during early spring when the ground is freezing at night and thawing during the day, broadcast seed at a third more than the normal rate over the slightly frozen crust. The freezing and thawing will work the seeds into the soil enough for germination when warm weather comes. You can also run a disk lightly over the thin sod later in spring when the ground is drier, to make grooves into which your broadcasted seed will fall and be covered when it rains. Sometimes burning dead grass in spring and then broadcasting seed on the bared, blackened surface results in a successful stand. But the best way to renovate a pasture is not to reseed it at all. Instead, manure it heavily, lime it, and then keep mowing whatever grows. In a couple of years, the native grasses and legumes that are already there, though depressed by over-grazing or lack of fertility, will volunteer—especially bluegrass and white clover.

Which weeds are harmful and must be extirpated from the pasture? I've mentioned white thorn. Burdock is another. Animals won't graze it hard enough to control it and the burrs ruin wool and mohair. Every area has a number of plants the livestock won't eat, like ironweed and bull thistle, and these should be mowed or hoed before they go to seed. White snakeroot is totally poisonous, and milk from a cow that has eaten white snakeroot can be poisonous. Fortunately this weed is found in woods and wooded fencerows, and then only rarely, and has not been a problem since pioneer days when animals grazed frequently in woodland.

Which plants do the animals like the most and therefore eat first? The clovers. White clover most of all. Young bluegrass is another favorite. That is why I try to maintain white clover and bluegrass as my main pasture plants (see below).

Also on this level of management, you need to know ecological details like where the meadowlarks and bobolinks are nesting, and plan to graze that portion during the nesting season in late spring instead of making hay off of it. The mower might destroy the nestlings and perhaps even the mother birds.

You will also want to resist the urge to cut every bull thistle that thrusts its spiky head above the grass. Yes, they can reduce the grazing value of the pasture if allowed to spread. But a couple will draw ground sparrows or song sparrows to nest under their prickly, protective arms. When you find a nest under a thistle, don't cut the thistle down; just clip off the seedheads. Like all annuals, bull thistles die in their second year anyway, after blooming.

On the other hand, spare no musk thistle, a very difficult weed to control, and one which fortunately has not yet made its way from the South to our farm.

Sourdock is another pesky weed in pastures. The animals will eat at it, but as it gets older, the plant becomes unpalatable. Sourdock sports a large seedhead, very pretty in dried flower arrangements, and will take over a pasture if not held in check. I pull them out by the roots, or if they won't pull, cut the taproot as for burdock, two inches below the ground level. A spade or shovel makes an easier tool for this job than a hoe. Likewise, mullein needs to be excised from the field in this way. Wild carrot is another bad pasture weed, but sheep eat it and keep it at tolerable levels.

Canadian thistle is our worst pasture weed because it spreads by both airborne seed and roots. It does not have a single taproot like bull thistles, sourdock, burdock, and mullein, which once cut below the soil, will not grow back again. But a strong sod, several mowings a year, and sheep nibblings keep Canadian thistle at bay. In recent years a disease that turns the tops of the thistles white, and a gray insect, the name of which I don't know either, have also reduced the vigor of these thistles.

Kentucky fescue can be a problem in pasture management. Although it makes a fair winter pasture, it grows so rampantly in summer that it tends to kill out other grasses. On a diet of only fescue, livestock may suffer from a disease called fescue foot. Some research indicates that pregnant mares on heavy fescue pasture may abort or not breed at all. After years of listening to the agricultural universities advise fescue,

many farmers are killing it with herbicides or plowing it up and reseeding better grasses. Don't plant the stuff, but if you have already done so,
don't panic. I keep a small section for winter pasture, and after ten years,
I am noticing that it thins out if not fertilized, and white clover and
bluegrass make inroads into the stand. Nature abhors not only vacuums
but monocultures.

A consciousness of the relationships between pasture plants and animals comes into play another way. Sheep, cows, donkeys, and horses do
not all graze the same way or prefer the same plants. We learned that our
horse, in early spring, would eat the new growth of heavy, coarse swamp
grass that was invading the pasture next to the creek. The other animals
shunned swamp grass at all times. Donkeys, I am told, will eat musk
thistles. Goats will eat almost anything, but not tin cans, and are perhaps the only animal that can control white thorn at least when the pest
is only a seedling.

Sheep and cows together will make more efficient use of the pasture
than either species alone, contrary to what western movie cowboys used
to say. An old tradition in stock farming holds that a pasture that can
carry a cow per acre will also carry a sheep and her lamb(s) because the
sheep will eat herbiage the cow won't, and vice versa. I believe that the
small cottage farmer like myself would do fine combining cows and
sheep, which is a more practical form of intensifying grazing than trying
to increase the number of one or the other species. Since my emphasis is
on sheep, I stock a full complement of sheep which in this area is considered to be about five sheep per acre, and then I add one cow for every
four acres. If time proves my little pastures will carry a bit more livestock
per acre, I will back off on the sheep to four per acre and add another
beef animal. Or maybe a donkey.

The donkey would serve mainly to guard the sheep from coyotes
and dogs, which brings up another uncanny pasture relationship: the
one between wild animals and tame. When the riding horse we used to
have gave birth to a colt, mother mare decided that stray dogs were not
entitled to democratic rights in our pasture and she viciously chased
them away. She must have kept the coyotes away, too, because we had
no problems as long as she reigned over the meadow. Evidently she
taught her snide regard for the dog kingdom to Betsy, our cow, because

after Betsy started having calves, she would also run dogs out of the pasture, and I presume coyotes. After a while, the sheep, who are as smart as any of us when necessary, would, at the first hint of a strange canine in the field, gather around the mare or the cow, even standing right under them for protection.

I noted in chapter 4 how grazing chickens scatter the piles of cow dung, effectively saving the grazier from the job of dragging a harrow over the field to do the scattering, and how chickens also eat the worm eggs of sheep parasites to help control that problem. Chickens probably eat the eggs of face flies and attic flies too, a job formerly done by scarab beetles which, as lamented earlier, we no longer are blessed with in this part of Ohio.

The manure droppings of cows and horses have their special significance to the pasture, and not only as the latter's chief source of fertility. When cows or horses eat herbiage going to seed, the seeds, especially those of clovers, often pass intact through their bodies. Having gone through the digestive tract of the animals and then dropped to the ground in a protective covering of moist manure, these seeds sprout much quicker and much stronger than in the sod. In this way new patches of clover, in particular white clover, are started around the field. Sheep manure seldom serves in this capacity because sheep tend to digest fully the seeds they eat. But this fact can also be used in pasture management. Sheep grazing on weeds gone to seed (I am thinking particularly of Queen Anne's lace) will not spread the weeds.

There is one more set of important relationships that you must bear in mind. For the greatest efficiency, especially on a small farm, the meadows must be involved in a working partnership with the cultivated fields. These cultivated fields, in their own annual rotation of crops, can become at the right intervals in their rotations part of the rotational grazing scheme too, especially during drouths when emergency pasture is needed.

Blending the two kinds of rotation is tricky. I have my meadow land divided into five sections, and move the livestock from one to the other on a more or less regular basis depending on how fast the grass is growing. But when conditions warrant, I move the stock onto one or the other of the crop sections of corn, oats, wheat, and hay to give the

115

meadows some relief and a chance to grow, for example late in a dry summer. In August, I can turn the stock into those fields of clover that have regrown from the first cutting of hay and that would normally be used for a second cutting of hay. I have never had to do this but I could if squeezed hard enough by drouth. But in August after wheat and oat harvest, I often turn the animals onto the new clover that I interseeded with the grains and that is now growing strongly. The animals can stay there in August until they eat down the stand of clover, and the clover will still have time to regrow before winter and make a fine stand of hay the next year. In September and October, I regularly turn the stock into the cornfield after I have harvested the ears of corn from the standing stalks. The animals will eat, with surprising relish, the dried leaves off the stalks. In September, November, and December I can turn the animals into the hay fields from which I took a second cutting of hay in August. I will most assuredly do that on the hay field that is scheduled to be plowed and planted to corn the next year. As part of their grazing in the crop fields, the animals also eat weeds out of the fencerows around the fields, saving me a difficult job. And in November or December, I can pasture the new wheat, sown in September, without hurting it for next July's grain crop.

In short, by being able to move the animals from the meadow paddocks to the crop sections, point and counterpoint, I can usually avoid feeding them my precious winter hay in late summer drouths. In some years I do not have to feed hay until the middle of December, which is quite noteworthy for northern Ohio.

In July I will also move the animals for a week to the woodlot where I do not want any more brush or seedling trees to grow for the time being. This simple practice keeps brush from shading out wild flowers like dogtooth violet, Dutchmen's britches, Jack-in-the-pulpit, hepatica, rue anemone, trillium, wild geranium, and many others that grow in the spring. If allowed in the woods earlier, the sheep would eat the wildflowers before they matured, thereby eventually killing them.

Finally, there is a relationship between humans and their meadows beyond the reverie the two can engender. When the children are small, and eventually the grandchildren, some of the happiest family days on the farm are spent sledding down pasture hills. Meadows are also the best places to fly a kite. Once I outfitted a group of children with makeshift butterfly nets, and watching them skip over the meadow was even more delightful than watching fireflies sparkling above the grass in the dusks of July. In April, before the ball diamond uptown dries off, our softball team uses the biggest meadow as a practice field. My son and son-in-law also use it as a golf driving range. These values should go on the cottage farmer's computer spreadsheet, but how?

Major Pasture Plants

Listing the most important grasses and legumes for the graziers' purposes sounds like an altogether straightforward proposition. Instead, wherever two or more graziers gather together, expect a loud argument. This is because no two farms are alike, and certainly no two counties, states, or regions. I am going to list the plants that work best for me, and that I think are the best for a cottage farm in the Cornbelt, mid-Appalachian piedmont, and New England. Others may argue for their favorites, and I will try to mention some of them too.

Bluegrass. It seems to be a penchant with pasture specialists in agricultural colleges to knock bluegrass in favor of some higher-yielding, coarser grass. These other grasses come in and out of fashion, like hemlines, but bluegrass is the blue jeans of the turf world. Over the long haul, it is the most dependable, manageable, and permanent of the cool-season grasses. It needs to be limed, two tons to the acre, every five years or so, unless you are blessed with natural lime in your soil like that paradise of horse farm meadows around Lexington, Kentucky. Bluegrass makes a solid turf, one that sheep can walk on even in early spring thaw if necessary, and cows and horses can be turned on it as soon as the ground is reasonably dry and solid without making deep hoof prints. If mowed occasionally in addition to grazing so that it does not go heavily to seed, and if rain is plentiful, and if grazing animals have generously fertilized it with their manure, bluegrass will provide pasture all through the growing season except in the heat of late summer. In 1992, with continuous rains, the bluegrass never went through its customary rest period in August at all.

The other wonderful thing about bluegrass is that in the region

117

where we farm, it will volunteer. Take any field and start mowing it regularly and within two or three years, with lime and perhaps some fertilizer, bluegrass will slowly take over. This characteristic seems almost miraculous until you try to keep bluegrass *out* of a raspberry or strawberry patch and learn how tenaciously it can grow. Contrary to what is often said, succulent new-growth bluegrass has almost as much protein content as legumes. Wendell Berry, the original contrary farmer, tells me that cattle can be fattened on good Kentucky bluegrass pasture just about as readily as on corn—and he's talking about land that is too steep to cultivate and that formerly had eroded to near wasteland because it was cultivated.

When we bought our farm, the part of the acreage that would eventually be pastured had been haphazardly cultivated to corn and soybeans and then just let lay in some miscreant government subsidy program. It was growing rankly with giant ragweed above a soil surface that was otherwise mostly bare. I ran the disk lightly over the soil and the dead weed stalks, limed it, and then scattered ladino clover seed over the surface with a hand-held broadcast seeder. I now know I could have used red clover just as well but ladino will stand wet and slightly acid soil better than red. The stand of clover was wondrous to behold and as it slowly disappeared over two years, volunteer strains of bluegrass and other wild grasses took its place. In very early spring I then broadcast white clover and birdsfoot trefoil over the thin, dormant sod. Both grew. The trefoil eventually disappeared. The white clover established itself permanently with the bluegrass and with several other wild grasses I can't identify by name. If you sow bluegrass seed, fifteen pounds per acre is plenty, and ten pounds will do.

White clover (*Little Dutch*) grows in symbiotic relationship with bluegrass and, in many areas, volunteers with it without special seeding if the soil is limed. Like all legumes, white clover draws nitrogen from the air and fixes it in the soil. When soil nitrogen rates go up, the bluegrass responds with lush growth, crowding out the clover. When that nitrogen is depleted, the bluegrass languishes a little, and the white clover comes back fat and sassy. This ebb and flow will continue almost forever if mowing (and occasionally hand-weeding when necessary) keeps coarse weeds and tree seedlings from establishing themselves, and if the pasture is not overgrazed.

Where rain is plentiful, white clover will continue to grow through the hot, drier part of summer. Otherwise it will "rest" like the bluegrass and then come back vigorously in September, with the fall rains. White clover is the favorite herb of grazing animals; at least my animals eat it first and graze it the hardest. The best planting rate is two pounds per acre, if you have to sow it.

Alsike, birdsfoot trefoil, and other lesser-known legumes all volunteer in my pasture. Alsike clover is the variety farmers used to plant before they were able to get their fields well drained with tile. It is a persistent clover, but less vigorous and not as tall as red clover. Alsike has a pinkish white blossom, unlike the reddish-purple of red clover. Alsike grows in places that are too wet for the other clovers. It evidently reseeds itself; I planted it only once.

I also planted birdsfoot trefoil once, and it continued more or less vigorously for five years. To my surprise, the animals did not prefer the lacy, leafy trefoil to other clovers but that turned out to be an advantage. The trefoil had time to root down well before the cows and sheep decided to eat it.

Several other clovers are present naturally in our meadow. I am not sure of their identity but people say they "look like" hop clover, black medic, and subterranean clover. I have a peculiar attitude about things that come to me gratis from nature. I don't want to learn the name for fear the knowledge will somehow scare the gift away. A patch of purple vetch volunteered in one pasture and quickly spread. The books say it is not very palatable but my sheep have not read those books and gobbled it up. I think they like a change in diet just as we do.

Wild grasses. As there are various wild clovers volunteering in the pasture, so there are several species of grasses that flourish without my help. I don't know what they are—some foxtails and wild millets and bromes plus what I call buffalo grass, and many others that I can't identify. Some of these grasses are the "warm-season" kinds that stand the heat of summer better than bluegrass and keep the pastures "alive" when bluegrass is dormant. I'm inclined to rely on them (plus red clover) for summer grazing rather than trying to introduce prairie grasses from the drier west as other graziers do. They are planting Plains grasses like bluestem, switchgrass, and Eastern gama for summer grazing.

Red clover and alfalfa I grow mainly for hay, but they make good semi-permanent pasture. Alfalfa will last longer (five to seven years) and some varieties have been developed specifically for grazing, like Alfa-graze. Red clover will last two to three years if the new two-season varieties like Redland and Arlington are sown. They will reseed spottily, as will alfalfa, thus persisting in the pasture and providing feed in dry summer. I prefer red clover to alfalfa because it grows better on our heavy clay soils and it isn't bothered by alfalfa weevil. I habitually broadcast a light seeding of red clover over portions of the meadows in late winter to add to natural reseeding. The standard planting rate of both these legumes is eight to twelve pounds per acre.

When heavy stands of red clover or alfalfa are pastured, as in hay fields, be sure some grass is available to the animals. I sow timothy with these legumes for hay, and have grass borders around the plots in the annually cultivated fields. When livestock have both legumes and grasses to eat, they are not as liable to bloat. Bloating is especially a threat with lush alfalfa. I would not turn cows or sheep into lush first-cutting alfalfa or red clover. Second and third cuttings are usually safe if grass is present and you fill the animals up on hay before turning them on heavy stands. I've never had a case of bloat.

Timothy is the best of the coarser hay grasses in my opinion. Animals like it better than the orchardgrass which has so often replaced it. Whenever you plant red clover, plant timothy with it. I often seed it over the pastures where the bluegrass is thin. Its seedlings are so fine, you hardly notice them coming up through the sod, but invariably, in late summer of the second year after planting, or sometimes the first year, it seems to suddenly jump up and go to seed. A little will continue to volunteer every year and provide pasturage when bluegrass is waning in summer. Sow ten pounds per acre with clovers.

Orchardgrass produces a heavy growth of grass for pasture or hay, and a stand will last about five years before it gets spotty and clumpy.

The best way to use it is to make a hay crop of the spring flush growth, and summer pasture the regrowth. Where I started an orchardgrass pasture, bluegrass and white clover slowly took over as the orchardgrass diminished. Some orchardgrass still remains, in spotty clumps, and that's what I like. It adds summer grazing when the bluegrass and clover are dormant and the animals relish the orchardgrass seeds when they are in the milky stage even though at that time they will hardly eat the maturing stems. Where a stand of orchardgrass is heavy, if you don't make hay of the first tall growth, by all means mow it in late June after the stock has gotten its fill of milky seedheads. Then the grass will grow new, tastier leaves. Sow fifteen pounds per acre.

When mowing off aging pasture plants, don't think you are wasting forage. The regrowth will be more palatable and nutritious, and the old growth acts as rotting mulch to hold in soil moisture and increase organic matter.

Ladino clover is a substitute for red clover on soils that are a bit too wet or acid for red clover. It looks like a large version of white clover and is almost as palatable, but lasts only two years in a stand. Sow two pounds per acre.

Big Mistake is my name for *sweet clover*. Its value (debatable) is as a green manure cover crop to plow under where annual cultivation is contemplated on compacted, depleted soil. The trouble is that it tends to spread all over the farm and becomes a weed. It is not very palatable and can become toxic as hay. Use alfalfa instead to break up plow pan and to add organic matter.

Bromegrass is considered one of the better warm season grasses over much of the U.S., specifically for late summer grazing. I say that with my fingers crossed on the strength of others' experiences. I haven't tried it. It is difficult to get a stand established in my region. Sow ten to twelve pounds per acre.

Ryegrass works best as a winter cover crop in fields bared by annual cultivation. Not much place for it in meadows, although when renovating a pasture you can add a little ryegrass to your regular seed mixes and get a quick cover on the field for erosion control and grazing while the bluegrass and white clover or red clover are getting established. New varieties of ryegrass are being touted for palatability and staying power but I've never been in a situation where I needed them.

There are many other legitimate grasses and legumes (like lespidiza, bermudagrass, crimson clover, and bluestem) that have their places in southern and western pasture management and many regional books and government pamphlets to inform you about them. However, you would do better to ask some old farmer who has lived a lifetime in your area. Be careful about introducing strange plants. Bermudagrass makes a good southern coastal pasture but it is an absolute curse in a Kentucky lawn or garden. Kudzu was the great midsouth wonder plant of the 1940s that turned into the great midsouth scourge of the 1950s. Hybrid Sudan grasses produce enormous amounts of forage but there are problems with prussic acid and the stuff is neither permanent, nor, according to my cow, palatable—not the right choice for cottagers. There is a big push to introduce western prairie grasses (bluestem, Easter gama, Indian grass) into the eastern half of the country for late summer pasture, and that may turn out to be a good move, but I don't know how these grasses will react to our humid climate over the long haul, and I can get all-season pasture from the grasses and legumes that have proven themselves here.

Whenever interest in grassland farming resurfaces, a new set of wonder plants comes into vogue. You can read about the current offerings in farm magazines. Most farmers go through about three waves of wonder plants in a lifetime, and when they finally learn not to bite, it's retirement time.

Urban Meadows

You can enjoy a meadow even if the only space you have is a large yard in town. Managing a lawn to look like a meadow, and not like a wild jungle of noxious weeds that will get you in trouble with neighbors and the zoning board, is not easy. You need to think of your lawnmower the way an artist thinks of his brush. The lawn is your canvas. You are going to paint wildflowers in the grass. At first leave a couple of square yards unmowed and watch what happens. By the third year you might be very surprised. Allow a couple more patches to grow. Where wildflowers (or weeds) bloom early, the plants will mature by July and you can mow those patches into respectable, WASP lawn again. Next year the wildflowers will come back. Where fall blooming wildflowers appear, you can usually mow those areas *until* July, and the fall bloomers will then grow. So now you know how to keep one part of the lawn "presentable" for the first half the growing season, and another part "presentable" for the second half.

Then you can get really sophisticated at creating wildflower landscapes in your lawn with a mower. Sometimes mowing at a height of 5 inches will encourage several kinds of wildflowers to flourish while maintaining a semblance of Prussian neatness that so delights the Germanic mind. Usually, if you pursue this delightful madness, you will work out a staggered schedule of mowing: Plot A is mowed in July; Plot B in October; Plot C every other August; and so forth. I have gone from complex madness back to simplified madness. Our whole backyard, being shaded, is easy to manage for wildflowers because the shade weakens the growth of grass to the delight of spring wildflowers. From April until June, the yard is lush with blue, white, and yellow violets, spring beauties, winter aconites, forget-me-nots, snowbells, grape hyacinths, wild strawberry, Siberian squills, Grecian wildflowers, metensia and crocuses. These wildflowers bloom and fade before the grass does much of anything. By June the grass can be mowed back to "neatness" without harming the wildflowers and they return every spring, spreading farther across the lawn. You can have the most talked-about lawn in the neighborhood (yeah) by letting trees grow up so that the lawn is always *partially* shaded and then starting spring wildflowers all over it. Winter aconite and snowdrop are the best because they will be blooming as quickly as March snows fade away, and will be gone in time for you to mow as early as mid-May and thereby remain on good terms with Prussian neighbors.

In a few spots where June-blooming wildflowers such as blue-eyed grass, firepink, Jacob's ladder, and Virginia waterleaf occur, we do not mow until July. The rest of the year we dodge around nice specimens of Deptford pink, goldenrod, or even ragwort and heal-all. All this makes mowing a little interesting, and increases eye to hand reaction time, which comes in handy when some young brute hits a million-mile-per-hour line drive back at me in the pitcher's mound.

John Fichtner, a farmer and agricultural teacher in Mason County, West Virginia, has been telling me about how he and his wife Carolyn and their children make use of their meadow—land that formerly grew only a weak stand of broom sage. Their farm is a remarkable example of meadow farming and ecological husbandry. The family lives in an earth-sheltered home that Carolyn designed. On eighty acres, they graze Suffolk and Perendale sheep, Scottish Highland cattle, donkeys, goats, and turkeys. They also raise a few hogs, chickens, pigeons, ducks, geese, guinea hens, and Border collies, every animal occupying a special economic and ecological niche in the panoply of interconnected life forms. The donkeys, for example, guard against coyotes. The Muscovy ducks control flies by eating fly eggs and larvae in the animal droppings. The hogs root in the winter bedding of the livestock and break it up into compost. The guinea hens with their loud squawking at anything unusual make better watchdogs than dogs. The collies herd the sheep. And because of the variety of grazing animals, each with somewhat different preferences in what they eat, all of the various kinds of pasture plants are utilized.

The Fichtners sometimes rotate the livestock from paddock to paddock daily. Moreover, their animals are selectively bred to grow economically on pasture alone (requiring little or no grain means that the farm does not need to invest in expensive machinery for raising grains), and also bred for parasite resistance. Breeding, pasture rotation, and the use of herbs such as kelp and garlic allow the Fichtners to avoid using harsh chemical wormers.

"Our goal is to produce animal products profitably from grass while maintaining a proper holistic balance between humans, animals, and natural resources," says Fichtner. "We have to pretty much learn from our own experiments because this kind of farming has been mostly ignored by our agricultural universities and agribusiness."

CHAPTER 7

Groves of Trees to Live In

The groves were God's first temples.

William Cullen Bryant

If humans suddenly vanished from this land, in just fifty years the forest would take back its domain. I think of that often as I look out over the countryside of field and village from the edge of my woodlot, my little citadel of native forest in the land of everlasting cornfields. Even after fifty years, the courthouse and grain elevators and St. Peter's church steeple that I can see from my mailbox would surely still stick out above the tree tops. But for the most part, a traveler in 2043, hacking his way through the underbrush, would stumble upon the crumbling, moss-covered houses and stores, and feel no differently than an archeologist discovering a time-lost temple in the green jungles of Equador. When I plant trees, I smile a little, thinking how absurd, in a way, is my work. Nature would gladly plant the whole humid world to trees if we would only get out of the way.

There are enormous risks involved in imposing the psychology of blacktop on a forest ecology. When we separate our soil and climate from their natural role of growing trees, we also separate them from their ultimate source of fertility and ourselves from ultimate sustainability. We do know better, but knowledge is not enough to keep civilization after civilization from disappearing back into its degraded soil to be eventually covered again with trees—or if things have gone too bad, covered with desert or radioactive dust. In "Conquest of the Land through Seven Thousand Years," W.C. Lowdermilk's classic 1938 report on the effects of "civilization" on the earth, there is a photograph that shows one of the four groves of cedars still growing where the biblical "cedars of Lebanon" once flourished. Amidst a few ancient cedars on the grounds of a monastery, the photo reveals cedar seedlings growing in great abundance. Outside the stone wall that protects this green sanctuary from goats and humans, the land is a rocky moonscape of desolation. As Uncle Ade used to say, you couldn't raise an umbrella on land like that if you stood on a sack of fertilizer.

Lowdermilk found the same tragic condition on Chinese mountainsides, where the forest groves of Buddhist monasteries were the only living green in a jumble of rocks and ravines from which the soil had all eroded. An overpopulated society had cut every tree for fuel while its livestock grazed off every blade of grass for food. Below, on the flood plain of the Yellow River—China's Sorrow, as it used to be called—he observed how silt from the mountainsides continually filled the river bed, causing the water to rise and flood the rich farmland. With a stubbornness that defies belief, Chinese farmers built up dikes, century after century, to hold the river at bay, until the water level was higher than the farmlands around it. (The Mississippi River now flows past New Orleans between levees that also lift it higher than much of the city.) Periodically, floods broke through the levees and drowned the farmers. In 1856, 1877, and 1898, for example, the Yellow River flooded and killed *millions* of people on the densely populated plains on either side. This was, in fact how northern China controlled its population for a while. Let the river do the killing. Let war and famine help. Let us forget the future and the past; on lust let no restraint be cast. Not until the last decades of the twentieth century did China recognize that stabilizing population with birth control just might be a better way. Will the rest of the world realize? Not yet. China farmed for forty centuries with amazing productivity before it admitted that increasing food production merely increases birth rates.

Tree groves and cultivated fields are both essential to the agricultural survival of human civilization. Trees grow where rainfall is ample, the same places were crops can be grown without irrigation. Trees are big signposts that say: "Farmer, sink your plow here." But trees are also signposts of caution: "But beware. Erase us from this landscape and you too will be erased."

The popular image of a farm is a place of cultivated fields with, perhaps, a meadow in the middle. Various machines plant and harvest the crops, which are stored in various buildings and fed to animals kept in the same or other buildings, all further processed into food in kitchens and stored in cellars. But a farm in a sustainable society is a place of tree groves as well, with an array of tools and other equipment for harvesting, processing, and storing wood crops. (It was still traditional practice, until the piston engine infected farmers with a deadly strain of greed, to leave ten acres in woodland for every hundred acres of cropland—environmental tithing.) Wood crops become tools, fenceposts, gates, furniture, buildings, occasionally food in the form of fruit and nuts, and most of all fuel. A woodworking shop is as integral to the cottage farm as a chicken coop. A woodburning stove is the cottage's symbol of economic and therefore political freedom. If my hero and fellow contrary farmer Harlan Hubbard were alive today, he would say that the governmental regulators have been so strict on woodstove emissions—and even attempted in the 1970s to penalize woodstove users with higher utility rates for having the gall to use less electricity—because they want to discourage self-reliance. Self-reliant people never favor the taxes that allow governments to become totalitarian.

As the price of wood continues to soar, a tree grove may, acre for acre, bring in as much money as field crops on the cottage farm. The wood we heat with saves us about $500 a year in other fuel. Last year I made about $300 worth of gates from our lumber and from saplings thinned to make room for future timber trees. We sawed out about $2000 worth of boards that my son will turn into furniture, and we sold four veneer logs for $1000 each. We framed our woodworking shop with our own wood, and as much as we had available at the time went into our barns.

All this from two groves of about fourteen acres total. The smaller grove of four acres, where our house and barn sit, also provides range for the chickens and an annual two weeks worth of grazing for the sheep. In addition, we pick several meals of morel mushrooms from the woods each year, boil down a few pints of maple syrup, and gather all the hickory nuts and black walnuts we can use. The deer save some of the wild apples for us. The birds leave us a few wild black raspberries and all the elderberries and paw-paws we want. As Uncle Pete liked to say: "One paw-paw pie a year is enough."

Unlike steel, plastic, rubber, oil, coal, gas, aluminum, and so on, wood can be cheaply produced on any cottage farm and then shaped, bent, and joined with low-cost tools into a multitude of useful and necessary items. Thousands of amateur and professional woodworkers attest to that.

But above all, wood can stop the silly, disastrous power plays of oil magnates in the Mideast. There is not a house in America that could not save half its heating and electric bills with a combination of wood heat and solar energy. If government would pay farmers to maintain woodlands instead of paying them not to grow corn, wood could replace much (who knows how much?) of the oil used in this country. Wood converts to fuel alcohol more efficiently than corn does. Morton Fry, another magna cum laude contrary farmer in Pennsylvania, rigged his pickup with a gasifier and drove it all over the country on wood to advertise the many uses of the hybrid poplar trees he grows and sells. "Sure it worked," he once told me with a grin. "And I never had to worry about running out of gas. Just picked up a few downed tree branches whenever I passed a woodlot."

If we deny that we are a woodland culture we will fade slowly into oblivion. Those few who understand that fact inevitably become cottage farmers and wood craftspeople, and they plant groves of trees to live in. In effect, they establish little centers for the preservation of civilization during these dark ages of earth-plundering—very much as the monasteries did in those other Dark Ages.

Wood is a most versatile material, not only for those purposes for which it is customarily used today, but for many tools and utensils that we now make with plastic and metal. And this could become even more the case if known technology were utilized. During World War Two, DuPont developed a chemical method (too expensive to compete with plastic at present) that softens wood so that it can be twisted, bent, compressed, and manipulated as if it were a heated metal alloy. The wood retains its new shape and dries harder and stronger than it was naturally. A soft wood becomes as hard as sugar maple and maple as hard as ebony, Wheeler McMillen (another contrary farmer and onetime editor of *Farm Journal*) once told me. "Light colored pines take on the hues of cherry, the glamour of rosewood, or the depth of mahogany," he wrote in his fascinating 1946 book called *New Riches from the Soil.*

Even without any such treatment, wood once worked well in light airplane and glider frames, and for car bodies. Plastic toys and bowls are serviceable, but wooden ones could replace them in many homes. The Dutch won't part with their wooden shoes because their wooden shoes are practical and economical. Dogwood still makes arrow shafts as excellent as those that archeologists have found from pre-historic times. Ash makes a good hay fork. Fiberglass handles will never comfort the hands and arms as satisfactorily as ash or hickory. Softball players are convinced that aluminum bats drive the ball farther, but none of them playing today have ever used on a modern ball a bat turned out of the black heartwood of white ash. Osage orange, once promoted as a substitute for wire fencing (with its awful thorns, it will indeed make a cattle-tight fence, though the pruning to keep it that way is very laborious), has tensile strength greater than steel. That is why it was once the choice for bows. It is so hard and strong that it was used for wagon axles, wheel spokes, pulleys, tool handles, insulator pins, police billy clubs, door thresholds, paving blocks, railroad ties, and almost indestructible fenceposts. With its warm yellow-brown color, it could make dazzling parquet floors, though I don't know if it has ever been used that way.

Another unusual example, just to whet your imagination, is mesquite, which ranchers in the southwest have spent a fortune in herbicides vainly trying to kill. As lowly as this wood has appeared to be, applied human ingenuity has turned mesquite into what its aficionados call the "jewel of the Southwest." On ancient houses in Mexico, exposed mesquite lintels remain solid after four hundred years. Jim Lee, one of the contrariest cottagers of my acquaintance, and one of the most successful, makes stunningly beautiful boxes out of mesquite on his isolated homestead in Texas. Besides furniture and flooring, mesquite is being marketed as chips for barbecuing. Mesquite jelly is being produced commercially as a dessert food. In earlier times, mesquite beans were utilized in vast quantities as a cattle feed supplement.

There are many other common and more easily-worked woods for the cottage farmer to utilize (mesquite is twice as hard as oak). Jim and

Beverly Lee, and son and daughter Joel and Wendy, who call themselves The Hummers, use dead cedar that lumber companies years ago left behind as slash as the mainstay of their rural woodworking business. They have been known to generate a hundred thousand dollars a year from this "cottage farm crop."

Every northern Ohio farm used to have a grove of catalpa or black locust for fence posts and since cottage farmers today will almost always be husbandmen and in constant need of sturdy fences, such a grove would be practical for them. Of the two, catalpa is better because it is soft, so staples drive into it easily. The softness and yet durability of catalpa make it very desirable to woodcarvers. On the other hand, black locust produces a good honey crop for the bees besides being an enduring wood. Both catalpa and black locust would be ideal for outdoor furniture since they resist rot for many years. That would avoid the extravagance of depleting the rainforests of teak so that yuppy gardeners can brag about their teak garden benches which no one ever sits on.

A catalpa grove can be harvested indefinitely. When the first trunk is cut for posts, new trunks sprout from the roots and another crop is ready in seven to fifteen years. In England this practice has been called *coppicing* since ancient times. Ash is another tree that will endure coppicing for many years. The best ash handles for tools come from second or third growth coppiced wood.

Some of my fence posts came from my grandfather's catalpa grove (which was bulldozed away for corn after "farmers went crazy," as my father used to say) more than sixty years ago. Grandpaw Rall used the posts first. My uncle used them in a second fence. And when that fence was torn down, I inherited the posts for a third fence. As these posts get to thirty years of age and older, the bluebirds find hollow places in them that they can enlarge into nests without noticeably hastening the posts' slow decay. If you were to give a money value to a fence post that will last sixty years (no treated wood or metal post on the market will) and maybe make a bluebird apartment for thirty of these years, a catalpa post is surely worth three times the going rate, or $10 each anyway. An acre will hold about a hundred trees at twenty-by-twenty-foot spacings. To figure the number of trees per acre at different spacings, multiply the spacings, for instance twenty by twenty, and divide that number (400 in this case) into 43,560, the number of square feet in an acre. A hundred catalpa trees at twenty years of age should provide four split posts from each trunk, or four thousand per acre. At $10 each, that's $40,000 every twenty years, or $2000 per acre per year. You would think that someone would get smart and start a fence post and outdoor furniture business with a catalpa farm (selling the scraps to carvers) but by the time those of us so inclined learn about such things, we are "too soon old and too late smart."

Wild cherry and black walnut are much easier woods to work than oak and much more attractive in color and pattern of grain in my estimation. Walnut is very rot resistant and once was used for posts, sheeting, and siding (I know two old houses entirely sided in walnut) but is far too expensive for those purposes now. Walnut's high price—and that of cherry—is an accident (or mistake) of history, brought on by a human-made, very unnecessary scarcity. Both trees will grow and spread like weeds in the midwest if given any chance at all. There are thousands of acres of small creek-bottom fields in northern Ohio that flood so regularly that grain cropping on them is a losing proposition, but farmers go through the motions of farming these lowlands for the grain subsidy payments that can be drawn. How much better if the government would just pay the farmers to plant these fields to black walnut, for which that land is perfectly suited.

Maple is even more easily propagated than cherry or walnut, because it will grow readily in the shade of older trees. When old hardwood forests of the midwestern and eastern United States are allowed to regrow, after being overgrazed for years, the first new wave of trees in the shade of the remaining old giants is predominantly maple. If these are sugar maples rather than soft maples, that's not at all bad news because no northern tree has as much potential variation in attractive grain for woodworking as sugar maple—from tiger stripe to birdseye—and the wood is suitable for most construction purposes, especially for furniture and such specialties as flooring. Moreover, a grove of sugar maples can be, with the production of maple syrup, the most profitable acreage on the cottage farm. (Read *The Maple Sugar Book* by Scott and Helen Nearing, the two all-time, venerable gurus of the Contrary Farmer Revolution.)

The commercial Christmas tree market is usually in oversupply, but it is always fairly easy to sell a few trees a year from locally grown groves to local people. This is especially true when the rage for professionally clipped and shaped pine trees has subsided, and customers want a more naturally shaped tree with space between the branches for all the ornaments they feel compelled to buy. Most spruces and some firs grow in this fashion with little or no trimming, thus saving the labor that takes most of the potential profit out of the Christmas tree business. This is something else I got "too soon old and too late smart" about, although I am now planting a few blue and green spruces every year for retirement income.

Growing Christmas trees is a good instance of how cottagers should approach advice on woodland management, which is to say in the same cautious way they should view advice on grain crop production. Don't take literally everything the foresters tell you. Their advice is usually geared to the mainstream silvaculture business economy, which I am hoping you will be wise enough to avoid. You want just *a bit* of extra income, maybe as few as fifty Christmas trees a year at $20 each. Your principle goal is not all the profit the land will bear, but enJOYment, living in a grove of trees.

Whatever you decide to do with trees, do something. With us contrary farmers rests a holy possibility. Our so-called "economy" has no place in it for enterprises that will pay off only twenty, forty, or sixty years from now, or not at all. Only those who have found a way to extricate their lives, or at least a little part of their lives, from the enslavement of that economy can "afford" to plant groves of trees instead of corn and cotton.

Managing Your Woodlot

This bears repeating: Much of advice that you will read or are told as the proper way to manage forests is given with an eye to making the top dollar. That means also spending the top dollar in overhead, with the price of borrowed money determined by a banker who does not know the difference between a cord of hickory and a cord of poplar (about 10 million BTUs of difference), channeled through wholesaler and retailer, both tacking on their profit margin, and above all, with salaries and wages added in all along the route to market.

If you thin and prune the way your forester decrees, you *may* bring trees to harvest a few years sooner, or you *may* add a few dollars more to the value of each tree. Then again you may thin out one sapling in favor of another and then have the favored sapling die as happened to me once. If you follow industrial-inspired advice, you will spend time and hard work pruning low branches from saplings that in two more years will prune themselves—lower branches die because branches higher up the trunk deprive them of sunlight. The forest will, in almost every case, take care of itself quite well, as it did for millennia. I divided my home tree grove into two sections and maintained one under the forester's directions. The other I left alone. After fifteen years, I can't tell the difference: if anything, the dominant trees in the part left alone are larger than those in the section where I worked hard at pruning and thinning.

Rarely will you have to plant trees in an old established grove. The old trees and the birds and squirrels will do the planting. All you have to do is stand back and watch, and perhaps occasionally cut down less desirable or overpopulated species adjacent to the more desirable or desired species—"releasing" the latter to the sun, as foresters would say.

You can make more money by not allowing any trees to grow old, selling every tree as soon as it reaches harvestable age, with young trees coming along promptly to take the place of those removed. But followed slavishly, that rule will mean no hollow tree homes for owls, one of the most beneficial of wild birds, or wood ducks or woodpeckers or nuthatches, or flickers or squirrels or bees.

There is much clashing and clattering in the debate over clear-cutting vs. selective cutting. I suppose I should side with the selective cutters, and I do if the alternative is to clear-cut whole mountainsides, as greed has often dictated. However, on a small scale, such as a thirty-acre woodlot, you should try to practice a harmony between the two methods. (In all environmental debates, the truth almost always lies somewhere in the middle.) If you cut just one tree out of, say, every 10,000 square feet, a hardwood forest will inevitably fill up with shade-loving trees such as maple because the absence of that one harvested tree (unless it is a huge tree) does not let in enough sunlight to encourage sun-loving trees such as oak, hickory, ash, and walnut. So whenever possible, select three or four trees together to harvest, then move perhaps eight hundred feet away and take another cluster of three or four trees: a sort of tiny-scale form of clearcutting, done selectively.

The sun rules the forest. All the trees are in mortal combat to grab sunlight away from each other. A tree will reach out with an extra long branch if there is a ray of sunshine out there not being used. Or it will bend or twist beyond what you think its trunk strength can endure to get away from the shade of a bigger tree. If it can't get to light, it dies. Lower branches die because upper branches block the sun. Eventually brush dies out on the forest floor because the tall treetops allow no light to finger through the canopy. Therefore, when a big old tree has reached its maturity, and you can see a few dead limbs appearing in the top, you might as well harvest it, because in holding on to the grand old dying patriarch, you are ruining or delaying the new patriarchs to come.

But I can't make myself follow that advice. I consented to selling four big veneer white oaks last winter, all of them about one hundred and sixty to two hundred years old, only because there are other two-hundred-year-old trees nearby, unsuited for veneer. These will remain until they die—that's my compromise. Not in my day will I own trees like these again. Nor will my children or their children.

Actually there is a more practical reason to resist an unquestioning obedience to "proper" forestry procedure, except maybe in cases of veneer quality trees which are rare and too valuable to "waste." (I always put the word "waste" in quotes when talking about forests because nature wastes nothing. Dead trees rot into humus.) We have learned that allowing a tree to mature and die before harvesting it does not necessarily mean much wood is lost, despite what the forestry handbooks say. In most cases, hardwood trees first begin to rot high in the crown and/or at the base of the trunk. Invariably, the wind will blow them down when most of the trunk is still good lumber and most of the branches are still good stove wood. (Hickories and white elm are exceptions. After they die, they rot very fast even if they remain standing.) A timber buyer will turn his nose up at trees that are partially rotted, but good logs can be salvaged and sawn into lumber to be sold or used for one's own purposes. We hire the services of a custom bandsawyer who brings his rig right into our woods and charges reasonable prices. This way we avoid the cost and hassle of loading and hauling logs to the sawmill. More importantly we avoid the always dangerous task of cutting down big trees.

In caring for a tree grove, one has the option of mowing between the big trees in an effort to make the grove look like a park. Most suburbanites who buy a little woodlot out in the country to build a house in, do this. Better a park-like grove than no grove at all, I suppose, but how much wiser it would be to allow some seedlings to start growing to take the place of the big trees when they die.

Sheep will turn a woodlot into a neat park much more effortlessly than you and your mower. But wait until late June, after the spring wildflowers have matured so that grazing won't harm the next year's production of flowers. To make sure the grove sustains itself, fence off little areas with circles of woven wire fence (about fifty feet in diameter) where volunteer tree seedlings and brush can grow up inside, safe from the sheep. In about five years the seedlings will be tall enough so that the sheep can't destroy them. Then move your circles of fence to other areas. If you use a fence twelve feet tall, deer won't be able to get to the seedlings either.

The forestry handbooks say that pasturing woodland is not good for the trees. Done in the above way, sustainability is insured. As far as pasturing hurting trees for lumber and especially veneer, as the handbooks maintain, I suppose that is true if you put a large herd of cows or horses into a small woods where the animals may trample around the trees and harm roots. But sheep certainly, and a small number of small cows, do not hurt trees in my experience, if sensibly managed. My best proof is that the grove from which I recently sold veneer trees was pastured for at least half a century.

A second proof is that I have observed many times that trees out in pasture fields, under which sheep stand on hot days to take advantage of shade, grow very well indeed. The trampling over the root zone of the tree keeps competing grass from growing there and the heavy manuring invigorates the soil. If you want to maintain a really healthy, heavy-yielding apple tree, for example, plant a *standard* tree (not one with dwarfing rootstocks that grow weakly and provide little shade) in a pasture field, and after it grows up allow a flock of sheep to enjoy its shade.

In addition to fertilizing the tree, the sheep will eat all dropped fruit and that helps keep the trees healthy. The sheep eat the worms in the fruit too so the worms can't continue their life cycles. Nor can diseases over-winter on old fruit if there is no old fruit. I can't state it as a universal law—are there any universal laws?—but such apple trees produce fruit about 70 percent free of insect and disease damage, without spraying.

Orchards are tree groves, to be sure, but what I'm about to say now I wish someone would have told me years ago. If you are in the business of growing fruit, then of course soldierly-ranked blocks of fruit trees you must have. On the other hand, if you are a cottage farmer wishing to have fruit for your family and your animals, and plenty for juice, and maybe even a choice few bushels to sell, you may want to consider allowing your trees to bivouac anywhere there's ample sunlight.

For example, grow sun-loving fruit trees as forest-edge trees. In nature that's where the fruit-bearing trees are. They grow on forest edges and will not survive the shade of the deeper woods. Another advantage of forest edges is that the forest ameliorates the temperature on frosty nights a little, perhaps that one or two degrees that spell the difference between fruit buds freezing and not freezing.

Fruit trees, by the same token, are excellent trees to grow in fencerows where in addition to plenty of light, the grazing animals are handy for eating up the drops and surplus fruit.

But the main reason for scattering fruit trees out along forest edges and fencerows is that these trees are, in my experience, harmed less by insect predation than the ones clustered in the more formal orchard. Even in the latter case, I now have a great mixture of trees, nuts and fruits together, plum and walnut, pear and hickory, mulberry and hazelnut, and apple and peach and even mulberry and juneberry. Rarely do two of the same kind stand next to each other. As apple trees that I planted first, following conventional advice, die from scab because of not spraying them, I replace them with scab-immune varieties. There are about six varieties now on the market, and Liberty is very good. The apple trees in the fencerows and along the woods are all seedlings or wildlings that have demonstrated some resistance to disease and insects or they wouldn't have survived since Johnny Appleseed days. If I were to live long enough, I think that on a twenty-acre place like mine I could

find spots for about a hundred standard-sized fruit trees that would produce a 50 percent marketable crop (say five bushels per tree at $5 a
bushel or $2500, not bad pocket money). The other 50 percent would
be for home eating, cider, pies, sauce, and lots for the animals. A glass of
cider a day is a wonderful aid to bowel health.

Starting a Grove of Trees

If new woodlands, much needed in some parts of the country, are to
flourish, cottage farmers will have to establish them. No one else believes
they can afford to wait forty to sixty years, if ever, to get a financial return. To start a tree grove where only grass or cultivated crops existed
before is easier than you think. You do not even have to plant trees if
there are woodlands nearby, although I will try to explain, below, why I
think you should plant a few. If you just stand back and wait patiently,
you can watch nature stage a play in about six acts that is alone worth
the price of letting the land "go back to nature." (Why does no one ever
say, "go forward to nature?") You will die before the play is over, of
course, but that is true of the whole drama of nature. The final curtain
call never comes.

Act One is the Takeover of the Tall Weeds. Queen Anne's lace, thistles, dock, ragweed, milkweed, ironweed, and many others will shade
out the grass, or grow in riotous density where a year earlier corn might
have stood in its marching-band rows. This world of weeds is paradise
for many birds: mourning doves in search of weed seeds, goldfinches
after thistle seed and down, and a whole choir of songbirds after poke-
berries. To neatnik humans, the field has turned into a terrible mess. To
wildlife and artist it has become feast for stomach and eye.

Within two or three years, Act Two finds the fruit-bearing brambles
and thorn trees coming on stage, sprouting from seeds brought to the
field by birds and animals. The brambles and brush soon shade out the
weeds. Most of the plant actors in this scene have the ability to spread
not only by seed but by root. Vines, especially poison ivy, Virginia
creeper, bittersweet, and wild grape also slip into the scenery, winding
and twisting through the brush. Grain farmers with a Prussian reverence
for neatness found Act One annoying and Act Two unspeakably repul-

sive. If this is one of their hillsides too steep to plow, they will mow the "mess" into the submission of grass even though they have no animals to graze. They thus commit themselves to a lifetime of such mowing, whereas, if they were patient, the "mess" would eventually turn into a grove of trees. Cottage farmers, however, delighted with the theater of nature, only brush-cut pathways through the brush jungle so that they can have a front-row view to all the action. These alleys become the logging roads of the future and should be laid out with that thought in mind.

Of action there is plenty because in Act Two there appears the greatest variety of the fruit and seed-loving birds that reforestation will ever bring. The raccoons and opossums are plentiful now, too, because they love blackberries. So do box turtles. Groundhogs, rabbits, and foxes go on population binges. Dominating all are the deer, nipping on the new shoots of the brushy growth and finding places to hide where no human and few dogs will follow.

In a few more years the curtain rises on Act Three. The jungle of thorny brush and bush starts to wane in a way that seems puzzling to those who have not yet learned the power of the sun. Tree seedlings are starting to push up through the bramble brush and shade out the lower growth. Wild cherry is the darling of the theater at this time, usually managing to upstage box-elder, elm, red maple, sassafras, ash, mulberry, and eventually that terrible villain, white thorn. Wind moves the seed of ash, box-elder, elm, maple. Birds bring cherry, sassafras, and mulberry. Deer and raccoons plus birds spread the wild haw fruit of the white thorn which is a villain, after all, only to humans and their livestock pastures. Its thorny boughs are favorite nesting sites and perches for many birds.

In small stream valleys returning to woodland, floodwaters carry black walnuts and sycamore seedballs from existing trees to lodge on the floodplains and sprout. Groundhogs, deer, and sheep eat the sweet pods of honey locust and after the seeds pass through their bodies they sprout readily. Squirrels and bluejays and chipmunks accidentally drop nuts and acorns or bury them and forget, and more trees are born.

Theater-goers whose favorite stars are the meadow birds find Act Three disappointing because the meadowlark and bobolink and bluebird disappear now, unless there is a pasture field nearby. The fruit-loving birds diminish a little, too, as the raspberry, blackberry, and elderberry bushes begin to be shaded out by the growing trees.

In Act Four, twenty years from the beginning of natural reforestation, the native hardwood trees have fought their way above the brush and thorn trees and begun the slow, patient work of shading out all the brushy growth beneath them, even the scourge of multiflora rose and the plague of autumn olive (the former once championed by the agricultural college experts as "living fences" and good habitat for wildlife, and the second still championed, unfortunately). The white thorns and wild plums will put up a good fight, but inevitably they dwindle too. The pathways don't need to be mowed anymore because the shade of the trees keeps them open.

By Act Five, in the grove's thirtieth year, the plants and animals of the early brush years have moved to the woodlot's edges where the bramble berries and haws and wild plums and bittersweet that the wildlife once fed on in abundance still find enough sunlight to grow food for a few. In the lighter shade just a bit further into the forest from the edge, the thrushes make homes in young tree forks, and the nuthatches and downy woodpeckers hollow out rotted knotholes for nests, while both peruse woodland and woodland edge for food. Also now, in spring and fall, you will see greater concentrations of the migrating warblers, the little jewels of orange, red, yellow, gold, blue, chestnut, white, and black that are my spring delight.

By Act Six the grove is fifty years old and though not completely matured to its climax of oak and hickory (that can take two hundred years), it is well on its way. The wild cherry trees have almost matured and except on the most favored, protected northern slopes will soon begin to lose vigor or blow down. Except on the deepest loams, the walnuts are reaching their limit of vigorous growth, too. Deep in the grove, the primeval silence of the ancient forest begins to return. The only bird chattering comes from high in the canopy where the jays and crows and squirrels and scarlet tanagers and hawks live in their own edge to the sun. Lower down, the owls and bats reign, keepers of the dark, owing no allegiance to the sun. On the grove floor grow fungi, some to eat, like morels, and some to awe the night-time wanderer with their phosphorescent luminosity. Wildflowers bloom in the spring sun before the trees fully leaf out or, like the mysterious Indian pipe, grow in the deep shade without chlorophyll.

Over centuries, the trees continue to grow. When the large ones finally topple, letting a shaft of light to the forest floor, seedlings spring up, and eventually, the one with most access to sun and to soil nutrients takes the old one's place. Forest fires and tornadoes open up large acreages to the sun, and then nature's theater enacts again its play of reforestation.

Return now to Act One, and become director and choreographer of the play. In Act One, all plants start out with the same amount of sunshine: there is no tall growth to shade new growth. Although eventually the field will become a hardwood forest on its own, you can often speed up that eventuality significantly by planting seeds of the species you wish to end up with, without going through the whole wave of thorny trees and brush. This seeding will of course be necessary if no mother trees are within two thousand feet of the new woods-to-be. Select the trees that are native to your area. Black walnut and white oak, currently our two most valuable hardwoods, are two species especially important to get started at once, along with hickory if you want good nuts to eat, since they must get above the brush right away and receive full sun. Plant the black walnut seed and hickory (or pecan in southern climates) in the low, black ground and the white oak acorns on the higher ground. Choose seed from old trees that have good-tasting, easy-cracking nuts in the case of walnut and hickory, the sweetest acorns in case of white oak, the least bitter cherries in case of wild cherry, and the sweetest sap in case of sugar maple. Choose a late fall day when the ground is soft. Carry a bucket of seeds. Drop one and sock your heel on it, driving it partially into the sod or dirt. And so proceed. Just getting the seed firmly in contact with the dirt is enough, although if it makes you feel better, you can cover the seeds completely with dirt. Actually, just scattering the seeds on top of the ground is sufficient to get some of them sprouted. Walnuts can be planted unhulled with a potato planter. One cottage nurseryman makes a furrow with a plow, drops the nuts or acorns in, and then plows the furrow shut. Rodents, especially chipmunks, will eat some of the acorns sometimes, so planting heavily and three to four inches deep deters them. If I were starting out with a bare field, I'd broadcast a bushel per acre and disk them in as if I were planting oats. That would discourage chipmunk predation, as well. If you want to get really professional or need to plant a fairly large area, the Truax Company in Minneapolis, Minnesota sells a regular acorn and nut planter.

Windblown and bird-scattered tree seeds will take care of themselves if any mother trees are around. Otherwise, you can gather the windblowns and scatter them over the field yourself.

Conifers are not native to our area, and I am not sure if my general rules of forest care apply to them. If we want evergreen trees, we must plant nursery stock from some other area. If left unattended, conifers here will be overwhelmed by the hardwoods that volunteer around them. Where we wish to have a grove of evergreens, we manage them as we would Christmas trees until they are twenty feet tall. Even so, the hardwoods will eventually grow up and shade the evergreens out. Given a few centuries and a few forest fires, oak, our climax tree in this part of Ohio, will dominate all.

The Art and Joy of Woodcutting

Cutting down a large tree should be an act charged with ritual: Candles burning, incense smoking, plump bishops in high hats holding forth in stentorian prayer. A tree that has experienced two centuries or more of life on earth deserves that kind of respect. I speak not so much of the godly spirit in it that humans of all times have sensed in great trees; I mean its wonderful accomplishment. I stand before my two-hundred-year-old white oak worshipping its feat of survival. Hundreds of other trees have competed against it for the sun and lost. Hundreds more around it succumbed to wind, disease, lightning, fire. This is the white oak to save acorns from, and I do. I justify my felling it only because it is dying and by thinking how its wood will now become furniture at my son's hands, burnished and beloved by generations of humans. In ending the tree's green life, I release its woody soul to a sort of life everlasting.

But in cutting down trees, be not too enraptured with the spirits of the woods. Tend as well to re-examining your insurance policy. Tree cut-

141

ting is not just a confrontation with the tree's death. The woodcutter's life is in jeopardy, too. Cutting trees is a risky adventure. When ignorant people speak so foolishly of the boredom of the countryside, I tell them to go cut a tree down.

I try, as I've said, to wait for the wind to blow my big trees down, but I can't always delay that long. In which case, a few suggestions.

The most dangerous tree to cut in my experience is the hollow tree. There is no way to predict how it will react when you try to notch it on one side and then cut from the other in the usual way. The solution is easy. A hollow tree isn't worth cutting down. Leave it for the animals and birds and the bees. How do you tell if it's hollow? Strike it with a sledge hammer. A hollow tree resonates like a drum. Strike a solid tree for comparison. A hollow tree will usually also have bird and squirrel holes in it.

Also dangerous is the tree with a forked trunk. The lower the fork, the greater the danger. A contrary farmer friend of mine is paralyzed from the waist down because of such a tree. As he cut through the main trunk, one of the forks split off without warning and crushed his spine. If you must cut such a tree, stay as far away from the fork with your cut as possible and cut parallel to the fork, not perpendicular to it—a drawing is necessary to show this clearly. And bind the forked trunks together with a heave log chain. But these cautions are not fool-proof.

A tree that leans pronouncedly is dangerous to cut down, if over about six inches in diameter. The larger the diameter, the more danger. Novices believe this sort of tree is easy to fell because it already leans so far over. Not so. If you saw in from the back side of the lean, which seems logical, the trunk may without warning split straight up the bole after you have cut in four or more inches, and on a large tree, the cut end of the split can kick out and kill you. Needless to say, this ruins the log, too. Leaning trees need to be cut not with the lean or against it, but sideways to it, starting a little on the inside of the lean and coming around slowing, first on one side and then the other. I recommend paying an expert to cut it for you. I have seen real pros cut a tree leaning over a fence so that at the last second, the tree swivels around and falls parallel to the fence.

I have not mastered that trick, so on smaller trees that lean too much toward the fence to fell them the other way, I cut into the tree from the side opposite the way the tree leans at a height *above the top of the fence*. I do not notch the other side. I let the chain saw cut into the tree only slowly and stop immediately when the tree begins to sway. It slowly sags over, and the uncut part of the trunk acts as a hinge. The top of the tree hits the ground across the fence, but the log stays above the fence, still fastened slightly to the lower trunk. Then I cut up the tree into firewood from the downed top and work back towards the fence. This method works well only on green trees. An elm, standing dead for several years, will sometimes snap off from its high stump when it falls, and drop on the fence. But even so, if you are cutting high (that is, above the top of the fence) the tree will usually hit the ground first on the upper branches so its full weight will not smash down the fence too badly.

My best solution, in the problem cases mentioned above, is to make use of an expert timber cutter when you are selling a few trees. "I'm afraid to cut down that tree over there," I say in my most woebegone voice. "While you're here, would you mind showing me how it's done." That sort of approach. The expert will invariably do it for you for nothing.

Another gambit we try is taking advantage of the telephone company or electric utility. If the utilities have lines along your woodlot, they practically demand that you call them when a tree is to be cut that could fall on their wires. They have learned that too often we amateurs drop trees the wrong way. So again, we play the old "while you're here" game. Often the utility man will oblige and cut another problem tree down for us if it isn't too far away from the utility line.

In felling a big tree, it is safer to stand close to the trunk than farther away. If there are observers (spouse, kids, whoever) either have them stand near by the trunk where you are cutting, or way beyond the possible reach of the branches. If the tree doesn't fall where you planned, the expanse of where the top hits the ground can cover a spread of a hundred feet or more. A person standing out there has farther to run to escape the unanticipated direction of fall. At the trunk, one has only to take a couple of steps to get out of the way, but make sure the area around the trunk is clear of debris and brush and grape vines that might trip you when you back away or step around the falling trunk.

Once a big tree is on the ground, it can be more dangerous than ever, believe it or not. The reason is that the huge trunk now rests not on the ground but usually ten feet or so off the ground, held up there by the branches. You must study those "legs" well before you start sawing. Twice I have seen a huge log twist wickedly when a supporting leg was cut off on one side of the log, and a widow-maker branch on the other side flip over to where the woodcutter was standing with killing speed. What I do to avoid danger is first cut off all free branches, that is branches not holding up the trunk, to get them out of the way. Then I start cutting up the wood from the branch end of the tree, always keeping my body ahead of the trunk as I work down it. That way if the trunk does fall or flip over without warning, I am out of harm's way.

Or after you have cut away all free parts of the tree, you can pull the trunk over and off its leg braces and to the ground with a tractor—making sure the log chain is long enough so the tree falls clear of the tractor.

If you get very, very clever, you can actually keep the log suspended on legs (where it is easier to cut up into stove lengths) until you have only a small section of log left. But no one can tell you how to do that because each tree is different and each situation after each cut is different.

I seldom see in books directions on what to do when the tree one is felling gets hung up on another tree. I think that's because there are no such directions. You can hitch a chain to the trunk, and try to drag it far enough with tractor, winch, or horses to pull it off the tree it is leaning against. With nothing powerful enough to do that, I have disentangled a tree by sawing off portions of the truck until it comes off the other tree of its own accord. It is *very* important to saw upwards from the underside of the log; if you try to saw downwards, the log will pinch your chain saw so solidly that you will have to get another saw to free it. Also when you saw from the underside be careful that the tree, cut loose, does not slide or swing down and hit you. Trees are very heavy things. They don't just hit you; they kill you.

The veteran treecutter seldom has to face the problem of hung-up trees, because he can drop them where he wants to, so they fall clear of other trees. The tree is notched in the direction the cutter wants the tree to fall. Then he cuts from the other side toward and a little above the notch. He never really knows if the falling tree is going to follow the notch exactly—often the weight of the tree will want to pull it a little to the right or left. To overcome that, the tree can, sort of, be guided down by the way you approach the notch with your saw kerf, cutting from the other side. The trick and principle is this: if you cut closer to the notch on the left side, freeing the trunk more on that side, the tree will tend to tip and fall a little toward the right side of the notch. If you cut farther toward the notch on the right side, the tree will tend to tip toward the left.

Experience, as ever, is the only real teacher. And sometimes experience doesn't help. I cut a tree down recently that in falling brushed against a dead sapling, bending it over and then releasing it. Somehow the top of the sapling snapped from its trunk and whizzed right over my head at about a zillion miles per hour. As I say, farming is not only not boring, but the most exciting profession of all (and statistically second only to mining in danger).

After a while, you learn that despite all its noise, a chain saw is a delicate instrument. The pressure you exert on it should be very light. Sawing a log, the blade should be *lifted* up and in, going down over the side of the log opposite to the woodcutter, so that when the log is half way sawn through, the blade is nearly vertical in the log, with the handle straight up into the air above the log. Only then bring the handle down *gently* while the blade cuts horizontally through the remainder of the log on the near side. Let the blade eat its own way along: apply only the slightest pressure. Remember that a chain saw likes best to be sawing with the tip end moving towards the wood, not the butt end. That's why those teeth are on the butt end: to anchor the butt so it can act as a fulcrum while the blade moves into the wood with the tip end leading and moving downward.

When dropping a tree, place some short lengths of small logs along the path where the tree trunk will land. These small cross logs keep the trunk off the ground enough so that sawing it up can be accomplished with no danger of running the blade into the ground and dulling it. There's nothing like dirt, or sawing through ice-encrusted logs, to dull a blade quickly. Also, with the trunk off the ground a bit, you can more easily foresee and avoid pinching.

Four to five cords of hardwood should be enough to keep the average, well-insulated home warm through an Ohio winter: a little more farther north, less farther south. For that much wood, I find a conventional eight-pound splitting maul ideal. No need for a mechanical splitter. Knotty pieces or crotch pieces that won't split by hand we save for our big fireplace or for boiling down maple sap and lard. Splitting wood is anti-stress therapy for me as well as exercise. I also think it keeps my batting eye and arm muscles in tune over winter. When I swing the maul, I always aim at a very particular spot on the chunk I'm splitting—just as a batter should do when swinging at the ball.

In other words, splitting wood is more sport to me than work. It is also a kind of art akin to that of the gem cutter. Most sticks of wood are hand-splittable if you know where to strike with the maul. With easy-splitting wood like red oak, you can cut up a log piece like cutting up a pie. With harder-splitting wood, it is much easier to split off edge pieces, taking about three-inch splits as you go around the stick. Smaller crotch pieces will split if you strike *with* the crotch, not crossways to it. You can't split a crotch at the vortex of limb and trunk. Won't go, no way. Some woods, like red elm, will not split by hand. I've been assured, though never tried it, that if the elm is frozen, it will split. I assume, if that is true, that the wood would have to be green—still have considerable internal moisture content. My elm is always deadwood because the Dutch elm disease kills the trees as they reach a diameter of six to eight inches. At six inches or less, dead red elm can be burned without splitting. Larger pieces make great back logs for the fireplace. Because red elm is so tough, there must be better uses for it than burning (traditionally red elm was the choice for wagon bolsters) but I'm not smart enough to discover what they are.

The Sacred Sanctuary of the Woods

I cut wood because wood heat is the only kind that makes me feel warm. To get the same comfort from other sources I would have to set the thermostat at 90 degrees which neither I nor the earth can afford.

But just as important, I like to cut wood. I like to be in the woods in late fall, winter, and early spring. When I first said that to friends, in answer to their question about what was odd behavior to them, they did not believe me—and to tell the truth, I was not sure of it myself. To enjoy a little woodcutting, an occasional walk in the woods, yes. But to do it persistently every winter, year in and year out? Would I continue to enjoy it?

Time has not changed me. If anything, I look forward even more to my days in the woods, splitting logs into stove wood. For one thing, winter is rarely harsh in the woods. If the wind blows cold from the west, I retreat to the east side; if from the east, I go to the west. I am reminded of the words of a very old gardener whose company I cultivated because he had haunted my woods and creek valley seventy years before I first set foot on it and so he lengthened my memory of it to over a century.

"Winters were warmer in the old days," he once said to me. I looked at him in surprise since most old people like to amuse younger generations with stories of how much colder it used to be.

"They were?" I responded, since he would not go on, I knew, unless I invited him to do so.

"The wind never blew like it does now," he said. "Damn fools cut all the tree groves down. Turned us into plains people. It was just a damn sight warmer when you lived in a clearing in the woods. Or had a grove stretching westward of you from north to south. We slept in unheated rooms then. By God I wouldn't do that today."

I never hurry in the woods. That is part of the joy of it. Working for money, we must forever hurry. The slavocracy of a wage economy, Scott Nearing called it. Sitting on a log, watching a downy woodpecker hop over the wood I have just stacked, looking for grubs in the opened log chunks, I realize I like woodcutting because I can split when I want to split, and sit when I want to sit. I forget all the world's bosses who hover over me when I leave the woods, reminding me that I must work fast enough for their profit as well as my wage, or what is the use of paying me.

A great horned owl hoots from high in the big trees, hoots for a mate. He does not have to worry about the slavocracy of wages. He gives a hoot about worthwhile activities, like mating. He hoots the first inklings of spring. I note how the black raspberry canes glow purple in the fading sun, a purple that grows in intensity as winter wanes. Another whisper of spring. I must bestir myself a little now because nature, too, is an uncompromising boss. If the weather warms even a little now, though it is still February, the sap will begin to rise, and the time for felling trees will have passed.

When a large tree hits the ground, the earth trembles. I can feel it in my bones and my soul trembles in response, like a tuning fork. And the sound that accompanies the fall causes me to quiver: as the trunk rends from the stump, a groan issues from the protesting wood and crescendos to a high wail as the tree gains momentum in its fall. At the last, the very last, it is screaming a raw, shrieking death cry, followed suddenly by a breath-sucking, thunderous WHUMP as the trunk hits the ground. In the second ripple of sound waves shuddering out from the great fallen giant, I sense, almost audibly, a kind of static snapping from branch to branch through the standing trees nearby. I imagine invisible sparks, the soul of a tree disintegrating. The bird chatter, even the murmurs of wind high in the trees, hush in the electric air.

What I have heard is the roar of a dinosaur dying. In a few more generations, no one in this place will know the sounds that a two-hundred-year-old tree makes when it falls. Should a philosopher ask in those days whether a tree falling in the forest makes a noise if there is no one to hear it, the answer will be: "Not anymore."

CHAPTER 8

King Corn

If the corn turns out as good next year as it did this year, we'll all be ruined.

A neighbor speaking to the author about surplus corn problems in 1982

The cultivation of maize has probably sent more soil down the Mississippi River in the last century than natural erosion did in the preceding twenty. Clever compilers of statistics say that for every bushel of corn produced, five bushels of soil wash into lowlands, ditches, and streams or blow into the next county. The townships of the midwest spend millions of dollars annually to clean out the dirt that oozes into roadside ditches, and this is the least of the erosion cost that the earth must bear. Much of the soil is lost from farms owned by wealthy people, who rent their land to farmers who produce surplus corn on it subsidized by taxpayers. Perhaps we are quite demented; I see no other suitable explanation.

I know a farmer rich enough to have retired years ago, who instead filled every building on his several farms with corn while he waited fruitlessly for a year when prices would go up high enough to make a killing on sales. When he ran out of other space, he knocked a hole in the roof of an abandoned country school on his property and filled this, too. That scene, plus one other, of a hog standing in the doorway of what had once been a pretty country church, say more about the decline of rural life than anything I could write.

So why do I place a high priority on corn for the cottage farm?

In the first place, it is a very reliable crop. When contending with the vacillating moods of Old Bitch Nature, reliability is everything. That is why corn is produced in excess. Even a fool can grow it and many do. Whenever I used to worry, in drouth years, that we would go into winter without enough feed for the livestock, my Uncle Carl, now passed away, would always scold me gently: "Now, Gene, never in my lifetime (he was eighty by then) have we ever lost a corn crop *entirely.*" With Carl gone, I have assumed the job of neighborhood soothsayer. Since our combined memories encompass a century, I think I can conclude with all the fervent rectitude of science that corn is, indeed, a reliable grain. Not in a hundred years have we lost a crop entirely. A Cornbelt farmer is not going to starve. He can always whip up a bowl of mush.

I do not say that entirely in jest. Although this book aims to tell how my friends, neighbors, relatives, and I enjoy farming and rural life, ours is no gentrified bucolic retreat surrounded by white board fences. Out here on the ramparts we understand that we are talking about survival. Nature is ready and willing to accommodate us dead or alive and preferably the former. Root, hog, or die is the password to her sanctuaries. We sense a miracle every time a seed sprouts and grows to maturity because, considering all the possibilities of that not happening, fruition *is* almost a miracle.

Corn is the most "miraculous" of the foods we grow. In 1988, rain did not fall here from April 12 until July 17. The corn stalks looked like pineapple plants. But after July 17, when some good showers came, the shriveling ears and stunted stalks on the best ground leaped toward the sky and made a fair yield after all. My special strain of corn made its usual giant ears, but I could find on many of the bigger ears an indented ring around the middle, where they had begun to stop growing in the drouth and then, after the rain, expanded and put three more inches of cob and kernel on the end. Even poorer ground produced half a crop.

Then came 1991, the year of the record-breaking drouth. But this time we went into the growing season with the ground saturated from the floods of 1990, and so the corn, native to the land of the Aztec heat, just went right on and made an average yield without any appreciable rain at all. I observed during the driest of dry spells in August that after a night of heavy dew every corn plant held a thimble-full of dew water in each cupped vortex where leaf met stalk. I calculate that every plant bore about half a cup of dew water, and with 20,000 stalks per acre, that's over 600 gallons per acre in one night. Some of this water was absorbed into the plant and some ran down the stalk to make a little ring of wet soil at the base. Birds, bees, and other bugs sipped this free irrigation water on the plants, and I imagine the mycorrhizal fungi and other biotic soil life threw some wild drouth-time parties in the moist soil around each stalk.

The next year, 1992, was the wettest year on record here. Yet our corn crop was better than ever.

Corn has its insect enemies, like slugs, wireworm, corn root worm, and corn borer, but if grown in rotation with other crops and if proper mechanical seed bed preparation is done before planting and proper weed cultivation practiced after planting, these pests rarely become a problem. Diseases also lurk in the cornfields, ready to lash out if opportunity presents itself. But for the most part, corn, because of the resistance bred into it by nature and by scientists, is not much affected. This might change, however, as factory farmers fog the corn land with poisons whose effect on the health of soil microorganisms is hardly half-known.

A second reason I place a high priority on corn is that it can be conveniently and even pleasantly harvested by hand in fairly large amounts—say, up to ten acres by two people. This is not true of other grains because their ears are too tiny for efficient hand harvesting.

When I say that corn is capable of efficient hand harvesting, most of my neighbors, reckoning profit in terms of hundreds of acres of corn, think I'm nuts. But on the cottage farm, as I shall try to show, small acreages of corn are sufficient and hand harvesting them saves significant amounts of money. All you need for harvesting are two hands, a husking peg, and a pickup or wagon or cart in which to toss the ears.

Look at what we have done with our "labor saving" technology. Consider for example a 160-acre field of corn. Forty years ago, this field was a whole farm, with house and barns and animals and many kinds of crops and people working in the fields. The buildings, woodlot, and fences have now been bulldozed away, the animals transferred to confinement factories where they must be heavily medicated to prevent disease. The people have gone away too, most to the boredom of the assembly line or the hopelessness of no steady job at all. In the spring, a lone man, often hired at minimum wages, jockeys monster equipment over the 160 acres and in a couple of days plants the whole parcel. Before or after, a custom sprayer applies poisons. All summer the field is empty of activity accept for an occasional Indian artifact hunter (me) and the deer. Again in the fall monster harvesters and semi-trucks or grain wagons appear on the field, causing terrible compaction if the soil is wet, and whisk, whisk, whisk, the corn is gone. One such "farmer" I know, groggy with lack of sleep after forcing himself to keep his $100,000 harvester going into the night, accidentally ran over and killed his hired man, who had crawled under the machine to check a bearing.

Meanwhile the people who might have harvested that corn with communal, physical labor, get their exercise jogging along the road, risking being run over, radios plugged into their ears in an effort to alleviate their painful, panting boredom.

Compare that picture to a scene I recently observed. I had gone into an Amish cornfield to find out from its owner when he would have time to press and boil down my sorghum molasses. Half of the field had already been cut with binder and horses, and a group of men and boys were moving across the field, setting up the bundles strewn on the ground into shocks. I approached them through the standing corn and they could not see me. I was in the situation all writers yearn for: I was invisible. Were these farmers bent over in pain and boredom at their "backbreaking, tedious work?" No way. They were jabbering away in German to the tune of almost continuous laughter while the boys wrestled and played tag between the shocks. *They were having a party, working hard.*

One year I grew four acres of corn instead of my usual one and invited in other cottage farmers to harvest it by hand. That turned into a party, too, and everyone took home a pickup load of corn for their animals and for making cornbread. Imagine what kind of future could emerge if this mingling of work and play were repeated over the whole countryside.

Even if growing corn were the most grueling work imaginable, and uneconomical to boot, I would still put it at the top of the list of necessities, for two other reasons: sweet corn and popcorn. Farming without raising and eating sweet corn ten minutes from the patch is like living out a lifetime as a virgin.

Maize thrives in regions averaging 35 inches of rainfall with 18 to

22 inches of that coming during the growing season of 90 to 120 days, and where summers are warm and long. But there are ancient varieties that grow in hot deserts of Mexico and high cool mountain ranges of the Andes. The Navajo cottage farmers and others have popularized some of these corns and built successful small businesses with their blue cornmeal. Dr. Richard Pratt, an agronomist at Ohio State, is trying to transfer the drouth resistance in these desert corns to regular Cornbelt strains.

Geneticists have developed short season hybrids to push the boundaries of the Cornbelt further north. Irrigation extends these boundaries farther into the West but the era of large-scale irrigated corn farms may be ending as ground water supplies decrease or become too expensive to pump out. In these drier areas, sustainability should dictate that other grains be grown commercially, including grain sorghum, a relative of corn, on the southern Plains, or barley in Idaho and Montana (more on these grains in chapter 10). Or perhaps scientists like Dr. Pratt will develop corn varieties for the Plains that won't require irrigation but will yield enough so that farmers can continue to stay in debt buying hundred-thousand-dollar combines that end up in demolition derbies.

But if I am so enthusiastic about raising animals on pasture alone, why do I consider cultivated grains at all, let alone corn? My answer: You cannot trust the weather to cooperate. Never put all your eggs in one basket. Pasture can be the sole source of animal feed, as the buffalo proved, but the buffalo migrated in winter and did not have to worry about superhighways. Cornell University has demonstrated that livestock in the north can be overwintered on pasture, even with lots of snow on the ground, but when you read the fine print you learn that, yes, the animals were provided with a little supplemental grain and hay when the weather got really tough. Animals will need a little grain in drouthy summers, too. A LITTLE is what the cottage farm provides.

Moreover the success of the cottage farm depends on biodiversity. The more species of life on hand, the greater the chance of survivability for them all. As Diane Ackerman says in her recent wonderful *New Yorker* essay on entomologist Thomas Eisner (August 17, 1992): "Variety is not just the spice of life but the indispensable ingredient." The contrary farmer lives by that dictum. If nature, in her contrariness, dries up the pasture grass, I will feed corn. If she hurls hail on the corn, I will feed oats. If she blows down the oats, I will feed wheat. If she floods out the wheat, the grass will be lush. A conductor can't depend on violins alone, but rather on a whole orchestra. And so must the husbandman.

Preparing Land for Corn

I have to laugh at the earnest garden and farm books (some of them my own from the 1970s, in my more innocent years) that propose to tell the poor reader exactly how to grow everything. The inevitable conclusion the beginner reaches is that there is one and only one right path to horticultural or agricultural correctness and that there are experts who really do know what path this is.

In practicing sustainable farming, corn, which is a grass, should be planted on land where some legume was growing the preceding year. That's about as far as I want to go in laying down inviolable rules, and even that one does not have to be slavishly adhered to. The ancient Romans understood this rule (see Virgil's *Georgics*; he got it from the ancient Greeks) and nothing science has done since has tempted me to seriously question the tradition. Which legume to use depends on your preferences if not your climate. I think the best one on my farm is red clover which makes me a contrary farmer once again: the experts all insist alfalfa is better but I stick with red clover because it stands humid weather a little better and is not bothered by the alfalfa weevil.

In the fall, in preparation for corn the next year, I plow the field that has been in red clover for two years. Yes, *plow*. And yes, in the fall. I could use a heavy disk and field cultivator or chisel plow to incorporate the clover for green manure but these implements are much more expensive than my old plow. I could plow in the spring, and sometimes do, but our Ohio clay soils are easier to work into a spring seedbed after fall plowing than after spring plowing.

Or I could spray poison weedkillers and plant into the dead clover with a "no-till" or "zone-till" planter that does not require working the soil into a fine seedbed beforehand. But I don't intend to pay $15 or more an acre for weed poisons if I can help it, nor do I need to spend $15,000 for a no-till planter and $60,000 for a tractor big enough to handle it when a $40 hand-pushed garden planter works just fine for me.

Nor do I want to be forced to spend $10 per acre or more on poisons to kill slugs and other insect pests which sometimes proliferate in undisturbed, no-tilled soils. As the slugs are killed, so also can be millions of earthworms and thousands of birds who eat the poisoned bait, as happened in eastern Ohio a few years ago.

No-till farming—that is, killing with poisons all plant life that might compete with the crop plants and then planting with special drills into the undisturbed soil—is being touted by the chemical companies and the government as the only practical (read, short-term, profitable) way to avoid erosion. No-till or chemical tillage does control erosion fairly well on fairly level to gently rolling land, but there are other less costly ways for the small farm to do that. The most vicious smear that the chemical companies brush on the face of good traditional conservation farming is the notion that no-till is the only way to stop erosion and that only no-till leaves a residue of plant material on the surface, which is ridiculous. You have the same amount of residue to work with no matter what method you use to grow the crop. What no-till farming really amounts to is a convenient way for large farms to keep on expanding in size while avoiding the severest kind of erosion that used to be common on careless farms. But erosion can still be bad on steep no-tilled slopes. And since no-till has only been practiced a short time, there is no way to tell whether it will remain a sustainable technique. In fact there is much evidence that weeds are developing immunity to herbicides.

I am a champion of the small plow and small tandem disk for small farms. I have the company of good contrary farmers: David and Dave, the first a horse farmer and the second a tractor farmer. They are both respected for their agricultural skills in their communities and are imbued with a high degree of ecological sensitivity. Like me, they are considered contrary by the Soil Conservation Service which has gone all out in favor of chemical-soak, no-till farming, and all but outlawed the plow on hilly land: If you want to get your government subsidies you can't plow what the SCS deems to be "highly erodible land." You can, however, rotate those very same acres from corn to soybeans year after year, spray the soil with weedkillers so that it is always bare except for the crop "residue" (the dead stalks after harvest), and watch this soil erode just about as badly, on sloping land, as it did with poor plowing methods. Just because you have residue on your soil surface does not mean you will not have erosion. On a steep, "no-tilled" field in Knox County, Ohio, a few years ago, I watched a gullywasher sweep away all that corn stalk residue which in modern chemical conservation parlance is supposed to stop erosion. Running water piled that "residue" up against a fence at the bottom of the hill with such force that the fence was knocked flat on the ground.

In my farming, I plow only sod land—that is, fields that were previously growing hay or pasture. That means any particular field is rarely plowed more than once every four years. Because of the legume roots, the infrequent use of the plow, and the applications of animal manure and green legumes incorporated by the plow (green manure), this soil stays loose and friable and I can plow it with a small tractor (35-horse-power) and a two-bottom (two plowshares) plow. David can plow the same way with a team of horses. As he points out, this is the only way a small farmer can cultivate soil efficiently. Alternative equipment to the plow and the small, tandem disk—standard chisel plows, offset disks, field cultivators, subsoilers, no-till planters—require much more horse-power. Some of us contrary farmers suspect the de facto prohibition against plows is just another ploy to help the big-iron ground hogs run us out of business.

My ancient two-bottom plow (I got it used for $10 and a pickup load of firewood eighteen years ago) doesn't turn the sod over completely, but sets the furrow up on its edge at a somewhat vertical slant. At first I was annoyed by that, since plow champions of yesteryear taught us to bury all that green manure from sight. The strip of green that stuck out at the top of every furrow my plow turned looked messy and God knows how my German forebears hated messiness. Then I learned from David and his horse plowing that the strip of grass peeking out of the ground was okay: it stops erosion as effectively as the vaunted surface "residue" of no-till farming. Furthermore, the plane of sod extending more or less vertically into the soil for six to eight inches along every furrow allows cascading rain to sink directly into the ground, cutting down tremendously on erosion. On my plowed sod, which in the first place is fairly level land, there is hardly any erosion at all. Even after I run over the surface with my old disk right ahead of planting, I am working up only about four inches of the soil into a fine seedbed. Underneath the fine dirt on top, the vertical shafts of rotting green manure are still there to draw down and hold rain. Therefore, in May and June, when the seedbed is exposed to the weather and when typically the really bad erosion occurs in this area (not over winter), heavy rains do not cause nearly as much harm as they do on hilly no-till ground that the USDA wants us to believe is erosion-free. In really disastrous storms, neither my way nor the no-till way will protect from erosion on hilly land. Hilly land should be in grassland most of the time. Grassland is the only true no-till farming.

Planting Corn

Field corn, sweet corn, popcorn, and "Indian" corn are all grown about the same way, but that way varies by region. The ancient, native American southwestern corns of the desert are traditionally grown in discrete hills, widely spaced, perhaps surrounded by pumpkins and squash, or beans that grow up the corn stalks. The kernels are planted as much as eighteen inches deep instead of the usual one- to three-inch depth in the humid Cornbelt. In contrast, the Mayans sometimes grew their corn on raised, irrigated beds and barely needed to cover the seed at all.

Commercial corn farmers with clay soils agonize over planting depth as well as time of planting. On any rainy day in early April, you can hear a conversation like the following if you sit around the coffee pot at the local feed mill.

Charlie: "It's too damn early to plant corn and everybody knows it. Ground's too cold." (He has already planted thirty acres but in case it doesn't germinate he doesn't want anyone to know.)

Hube: "Well I got two thousand acres I gotta get over, and if I don't start in April, I'll never get done in time to plant the other two thousand to beans."

Bob: "Don't bitch to me. If you weren't such a ground hog you could farm sanely and wait till the soil warms up."

Charlie: "How deep you plantin'?"

Hube: "Waaall, I started out at three quarters of an inch and then the weatherman said dry weather was ahead, so I went down to one and

157

a quarter inch, and then the weatherman said it was going to rain after all, so I went back up to an inch."

Charlie: "Hell, I'm just going to turn off the radio and put all the corn in at an inch and a half."

Bob: "And then it'll rain hail columbia and the ground will crust over and the corn'll leaf out underground and die.

Charlie: "I'm not working up a fine seed bed—that's what makes the crusting bad."

Hube: "Waaalll, then if it doesn't rain, that ground won't hold moisture and the corn won't sprout a'tall."

Bob: "A man's gotta be a damn fool to be a farmer."

Most of the time corn does just fine and will even make up for a judgmental mistake on the part of the farmer. When I was a novice, I tended to err more on the side of planting too deep in early spring, causing the corn kernels to rot in the cold ground; now I tend to plant too shallow, especially in early plantings when the soil has not warmed up fully. I tell myself, mistakenly, that I must plant very shallow, just under the soil surface, because the soil is too cold, even down an inch, for the kernels to sprout. My mistakes have taught me that it is better to plant at least an inch and a half, no matter how cold the ground temperature at planting time, in case dry weather does not provide moisture enough to germinate shallower plantings. Generally speaking, if the soil is dry enough to cultivate, it is warm enough to sprout corn at the inch-and-a-half to two-inch depth. That is the safest rule to follow.

The best lesson I have learned is not to be in a hurry. Farmers are just like gardeners. If someone in the neighborhood starts planting, everybody else thinks they have to rush out and take a crack at it, mud and all.

In the Cornbelt, we now plant corn in continuous rows although the old spacing in hills, three to four kernels to the hill with the hills 40 inches apart, still works quite well. (One for the borer, one for the crow / one for the cutworm and one to grow.) When sown continuously in the row rather than in hills, the kernels of hybrid corn are generally spaced about 6 to 8 inches apart. My open-pollinated (non-hybrid) corn does best at 12-inch spacings but can be planted a little closer than that.

Row width is arguable forever. The traditional width of 40 to 42 inches was dictated by the work horse, which needed that much room to fit between the rows during weed cultivation and harvest. The fact that corn is grown in rows at all was to accommodate beasts of burden, weeding tools, and harvest machines. Hybrid corns today are bred to be grown in rows 30 inches apart and occasionally in 20-inch rows, but 34- to 36-inch rows are more the norm in Ohio fields among farmers who intend to rely on mechanical cultivation of weeds and not herbicides alone. In gardens and beds, you can plant as densely as you have water, nutrients, and sunlight for.

A good example of small-scale but big league commercial corn production is the four acres one of the Ohio State Extension agronomists, Jay Johnson, has been growing continuously since 1989. He wanted to see how much corn would yield in Ohio if he used every available bit of modern agronomic know-how and focused it all on a very small field where he could exercise great hands-on care. He is averaging on the four acres about 230 bushels per acre, in a state where the average yield is about 120 bushels per acre and where 200-bushel yields, even with irrigation, are rare.

To get such high yields, Johnson uses a seeding rate of 35,000 kernels per acre or more (normal is 22,000 to 26,000 and I prefer 18,000 per acre with my open-pollinated corn). A plant population this dense requires very high fertility and hybrid varieties genotyped for dense plantings. He applied in the first six years nitrogen at the rate of between 400 and 500 pounds per acre, and in later years about 300 pounds per acre. Rates of phosphorus and potash he used at first ranged from 100 to 200 pounds per acre and 200 to 400 pounds per acre, respectively. In the last four years however, soil tests indicated that he did not need any added phosphorus and potash so he did not apply any.

What he learned was that despite continuously heavy applications of fertilizer and optimum moisture conditions, the corn yield leveled off after reaching the 230-bushel-per-acre level. He says that his rates of fertilizer were probably too high, and since continuing them did not continue to boost yield, he is now concentrating on seeing what rates are necessary to maintain yields of 200-plus bushels per acre. He also irrigates if necessary to maintain 18 to 22 inches of water over the growing season.

He says he did not keep track of his expenses, but obviously, if a farmer were to apply that much fertilizer, his costs would be higher than the norm (see the sample corn production budget below). I asked a seed corn salesman I know, who also farms, if he had ever grown 200-bushel corn. "Yep. Once. Never again," he replied, enjoying my mystified expression. "I made the 200-bushel club. 207 to be exact. But it cost me $3 a bushel to do it and corn was only worth $2.50 that year."

At Ohio University, botanist Ivan Smith grows corn on small beds using close spacings, hand weeding, supplemental irrigation, and biointensive nutritional programs that do not require any chemical fertilizers. He reports yields on small beds that projected to over 350 bushels per acre and says that conventional agribiz agronomists at Ohio State University wouldn't believe him.

For a more typical real farm example, Dave, near my place, averages 120 bushels per acre on his upland soil of fair fertility, and often gets 150 bushels per acre on his richer, creek-bottom fields. His farming philosophy has always been low-input production—moderate amounts of commercial fertilizer and herbicides, supplemented by clover incorporated for green manure and low-cost mechanical cultivation of weeds. He plants about fifty acres of corn now in semi-retirement but never planted more than a hundred. He spaces his rows 38 inches apart and uses seed plates in his four row, tractor-pulled planter that drop a kernel every 8½ inches. He plants hybrids, half early varieties and half late varieties. In 1992, with more than ample moisture, some of the corn yielded over 150 bushels per acre even on high ground, an excellent yield for his kind of soil.

The way I grow my corn is somewhat unique because I plant and harvest only one acre. My method could be expanded to at least fifty acres with small traditional equipment, in which case it would closely follow Dave's regimen but without the herbicides. My per-acre costs are lower than those of a commercial grower, but so is my yield. I grow open-pollinated corn which ordinarily has a yield potential of only 110 bushels per acre. I like to apply a ten- to fifteen-ton-per-acre application of manure right before plowing the legume under and often that is all the fertilizer I use. However, I have learned from experience that if the legume in the previous year(s) was cut for hay several times, the soil may be depleted of available potash. So the corn crop, planted in the spring, especially on thinner soil, will take good advantage of a moderate amount of additional potash, either as wood ashes, or a natural rock source of potassium, or some chemically refined fertilizer like potassium sulfate or muriate of potash. Potassium sulfate is considered preferable to muriate of potash by soil ecologists because the latter contains chlorine which may build up harmfully in the soil with heavier applications. What I actually do if manure is short is add nitrogen, phosphorus, and potash at about 200 pounds per acre in a 6-24-24 formulation. Those gobbledegook figures mean that for every 100 pounds of the formulation, I am applying 6 pounds of actual nitrogen, 24 pounds of actual phosphorus, and 24 of actual potash. (Compare that to Johnson's rates of 400 pounds, 200 pounds, 400 pounds.) I use this acidulated fertilizer rather than the raw rock sources of phosphorus and potash because it is cheaper and requires handling much less tonnage.

Oh, so you're not a certified organic grower, visitors sometimes snort, knowing that I have been an outspoken supporter of organic farming methods for thirty years. Nope, I'm not. For thirty years I've also said that a *moderate* amount of chemical fertilizer is not harmful to anything and is sometimes the most economical way to maintain fertility. I thought that the organic fraternity would eventually realize this. My garden is 100 percent organic, whatever that means anymore, and so is my orchard, though in the latter case I experimented many years to work out a way to make this practical. If a small amount of chemical fertilizer every fourth year on my crop fields makes me a cheater among the Chosen, that will prove in the long run to be the organic fraternity's problem, not mine.

The contortions that the National Organic Standards Board has been forced to go through in arriving at a list of materials that big time organic marketers agree should or should not be used in organic farming is evidence enough of the impossibility of defining "pure" organics to everyone's satisfaction. To effect a compromise, the NOSB has drawn up an interminably long and tedious list of materials, decreeing some to be okay, some to be used with certain restrictions, and some to be altogether prohibited.

How can a substance be restricted sometimes and sometimes not, and still be "organic"? So the Board conspired the status of "transitional" organic farmers—farmers on their way to becoming organic but not there yet. In some cases, these farmers can put "transitional organic" on their food or fiber labels, which is meaningless in terms of the product itself.

I think the NOSB has given itself an impossible goal: to try to administer an ideal from on high for thousands of farms in thousands of different situations. A *totally* sustainable agriculture might not even be possible no matter what known methods are used, given an ever-rising population. Marty Bender, who has made the study of sustainable agriculture his career at the Land Institute in Kansas, points out in a recent study that farmed soil will sustain itself in fertility without some form of added phosphorous and potassium only if it is not farmed but devoted to green manures half the time—for instance, seven years out of fourteen in tests conducted with dryland wheat. Since the sources of phosphorus and potash are finite, maybe the best we can do with rising populations is to slow down the environmental deterioration—delay the inevitable collapse until humans come to their senses and control their numbers voluntarily.

And to say that natural raw phosphate or potash derivatives are okay for sustainable agriculture and the fertilizers refined from them are anathema is an argument for which little proof exists outside the minds of organic purists. Besides, think of the increased air pollution, ozone build-up, transportation costs, and highway accidents that would occur if all of agriculture decided to rely only on far greater tonnages of the unrefined material.

I figure I am just as competent and justified in this situation to declare that a little chemical fertilizer is okay in good farming as the NOSB evidently feels justified and competent in saying that antibiotics, which I have not had to use, are sometimes okay in organic farming.

I also part company with the organic certification effort in its complete prohibition, no exceptions, against applying clean, pre-treated biosolids (sludge) on farm land. This prohibition is simply witch-hunting. I have covered the waste management field journalistically for nearly a decade, and have worked closely with the leading scientists in the field, including Harry Hoitink and Terry Logan at Ohio State,

Rufus Chaney at USDA, and John Walker at EPA to get the information on advances in handling biosolids to the public. These men, who have no axes to grind either way, insist that pre-treated biosolids today are among the safest of farm fertilizers if applied properly.

If anyone should be in favor of using biosolids in farming, it ought to be organic farmers, since this waste material is rich in organic matter. Marty Bender, commenting on the finiteness of sources of phosphorus and potassium for fertilizer, says that agriculture MUST turn to recycled city biosolids as one source for maintaining soil fertility. Gary Wegner, a wheat farmer in Washington state who is very sensitive to the ecological problems of agriculture, likes to point out that far from being harmful to farming, the minute quantities of so-called heavy metals (he thinks "lite" metals would be a more appropriate term) are precisely the trace minerals his soil needs.

Oren Long, a Kansas rancher, ecologist, and organic baby beef producer before he semi-retired, said to me this very morning on the phone: "If the National Organic Standards Board insists on barring all synthetic fertilizers, most farmers trying hard to achieve a sustainable agriculture that works will simply drop out of the organic movement. I don't care to be certified organic anyway; I just want to run an ecologically sane farm on the least amount of outside inputs as possible."

I throw down the gauntlet. I will match my corn (or any other food I produce) with any corn of any big food company who has muscled into the organic market and gotten certified. Measure both, sample both, diagnose both by the most advanced analysis available. If my food is not at least as pure of harmful chemicals or as full of nutritional content as the certified stuff, I will eat a football helmet and shut up forever.

The problem with straining gnats in organic farming is that it leads, sure enough, to swallowing camels or at least giant-sized oxymorons. I have listened to women condemn apple growers for using the chemical Alar, which has never been proven to hurt even one tiny baby, while in almost the same breath, the very same women support abortion as a means of routine birth control.

But I have more interesting things to do in the corn field than argue about gnat-straining organic farming certification. I am trying to grow a non-hybrid, Reid's yellow dent, open-pollinated corn that would be especially suited for hand harvest on small farms. Open-pollinated corn has the ability to produce much larger ears than current hybrid strains. Large-eared corn is more efficient for the small grower: I can husk one big ear much faster than two small ones. Furthermore, with open-pollinated corn, farmers can save their own seed instead of having to buy new seed every year as in the case of hybrid corn. Hybrid corn will return to its nubbly parent lines if replanted too often. Hybrid seed costs roughly $24 an acre (much more in 1993–1994, because of the great floods of the Midwest). In addition, there is some evidence (not yet conclusive) that open-pollinated corn is more palatable to animals than hybrid.

So the challenge presented itself. Could I, by selecting the largest and fattest ears from the sturdiest stalks every year for replanting (weak stalks are the bane of open-pollinated corn), develop a strain of strong-stalked corn with ears that contained twice the kernel weight of hybrid, thereby doubling the efficiency of hand harvesting?

Why aren't scientists doing this kind of work with open-pollinated corn? They do it with cereal grains which are almost all open-pollinated varieties. The answer is that corn is not homozygous the way that wheat and oats are. The seed does not repeat itself uniformly. When you plant all the kernels from a single ear of corn, many of the resulting plants are apt to be a little different from each other. Possessing that much heterosis, as a geneticist might say, corn is difficult to improve by selection but relatively easy to improve by crossbreeding. So hybridizing was born. Stalk strength and ear uniformity and increased yields soon followed.

But heterosis also means that corn has great potential for diversity, for producing the unexpected, and that intrigues me. How long an ear of corn is possible? Can I breed-in stalk strength by natural selection? When I asked Dr. Pratt if we could now have an improved open-pollinated variety just as good as our hybrid corn if that were a priority in research, he paused a little, knowing the political ramifications involved, and then answered: "Maybe."

It might take a long time and a lot of back-crossing to do it, which means lots of money. If done by the public land grant college scientists, who are the ones who normally maintain the banks of germ plasm (seed stock) from which new varieties are developed, it would also require quite a bit of courage: they would run into considerable opposition from the hybrid seed corn companies who of course are big donors to the land grant college coffers.

So, being contrary, I decided that if no one else would do it, I'd try.

Since then, instead of a tedious job, harvesting my corn by hand has become a most interesting pastime. Almost every ear is different, some high on the stalk, some low, some fat, some thin, some reddish, some almost white, some with red cobs, some with white, some with softer kernels, some tasting sweeter than others, some maturing late, some early, some easy to husk, some difficult. I save out about twenty of the biggest, fattest ears with the deepest kernels from the strongest stalks, shell the middle kernels out in the spring, mix them together, and plant them. Now, as I husk my way down each row, I keep thinking about the possibility that at any moment I might find the longest ear of corn in the world.

That might have happened in 1993, if I can believe a recent television show. My corn produced about a dozen ears that even when dried down to 15 percent moisture, measured over fourteen inches long. (As moisture decreases further, the ears will shrink a little more.) Quite a few ears measured over thirteen inches long, and twelve-inchers were common. This is corn that eighteen years ago, when I began my selective breeding, rarely produced an ear a foot long. I was encouraged, of course, but not too elated until one of those morning talk shows on TV featured a farmer who had with him what he said was the longest ear of corn in the world. It measured fourteen inches—not quite as long as several of mine.

Lots of people want to buy some of my strain but I am not yet satisfied with stalk strength. There has been much improvement in that regard, but I hope for more. A friend commented: "Gene, you are fighting a losing battle. Every time you get a stronger stalk, the ear gets bigger."

Caring for Corn

I have noted in chapter 3 that I plant corn with two little plastic push seeders bolted together to plant two rows at a time. Each seeder cost about $40. Many, many other kinds of new and used tractor and horse planters are available. Used ones at very reasonable prices sell regularly at farm sales. But in the planter department, the small farmer does not have to resort to old equipment if he can afford new. New planters can be purchased by the unit—one-row, two-row, three-row, whatever, and attached to a tool bar behind any three-point hitch tractor or horse fore-cart.

The art of controlling weeds is at the heart of successful farming. A low level of weeds can be tolerated, and is in fact helpful from the stand-point of biodiversity, but weeds are sort of like chickenpox: hard to have just a little.

Effective control of weeds in row crops like corn is a matter of good drainage and good timing, which sounds like a strange, unrelated obser-vation. At least it sounded strange to Pat and Joe even though they had both grown up on farms. Hungering to return to a small farm, they de-cided one year to turn a two-acre lot on the edge of the village in which they lived into an organic sweet corn farm to make some spare-time money. Tater-King Smith, so nicknamed because of the prodigious pota-toes he raised, tried to warn them.

"That field lays too wet," he told the young couple.

"Lays too wet?" Pat frowned, giving Joe a puzzled look.

"Yup. After hard rain it won't dry out until the Fourth of Joo-ly," the old man said. "Needs tile."

Pat and Joe considered that Tater was just putting on airs as old farmers love to do. They had everything figured out, they thought. They paid Tater King to plow the two acres, which he did while shaking his head and mumbling grouchily to himself, and figured that they could rely on a two-row garden seeder like mine, their garden tiller, and their hand-pushed wheelhoe to care for the crop. With successive plantings, they would have to handle no more than a quarter-acre a day and sel-dom that much.

They went ahead and made their first planting on May 2. That turned out to be the last day they were able to do anything on time. Rain fell on May 4 just as Joe was about to try out his new wheelhoe cultivator which he had discovered was just about as easy to push along as the garden tiller was to guide, especially up close to the row. Five days later when the tiny corn plants were peeking through the soil along with several hundred billion trillion weeds, the soil was finally dry enough to cultivate. But then rain came again. By the time the field dried out enough the second time, five more days later, the weeds were in full command, and it was slow going even with the tiller. Weeds in the rows could not now be buried by rolling dirt in on them. Only hand hoeing was effective, and with the soil now hard, and the weeds four inches tall or more, weeding became slow and exhausting work.

Had the field been properly drained, Pat and Joe would have had no problem controlling the weeds despite the rainy spring. The field would have been soft but dry enough to cultivate two days after the first rain, and with wheelhoe or tiller the soil would have worked up like applesauce, flowing into the rows to bury the still tiny weeds there while the weeds between the rows were being cut out of the ground. If a little hand hoeing were necessary, it would have been easy work in this soft mellow soil.

In the poorly drained field, the weeds overtook the first planting. The dirt was as hard as concrete when it did finally dry and the shovel cultivators Joe tried to use behind Tater's farm tractor turned up slabs as big as bricks which rolled over and dropped on the corn plants. Didn't much matter anyway because the weeds in the rows were as tall as the corn and could not be buried without burying the corn too. The same sad chain of events followed the second planting, which was delayed by wet soil until it was almost time to make the third planting. Therefore what corn Pat and Joe got to sell came all at once late in the season after most potential customers had eaten their fill from other farm stands.

"I believe we lost only about $50 an acre on the two acres," Joe said sarcastically to Pat at the end of the year.

"Lucky you didn't put out more than two acres," Tater King observed.

Ever after, Pat and Joe swore that growing organic corn was impossible because only with chemicals could they control weeds in wet weather.

In well-drained soil it is also possible before planting, and especially practical on larger acreages, to disk the plowed seedbed at least twice at intervals of about four days, in order to destroy the first two waves of sprouting weeds. This also prepares a clod-free seedbed that is easy to

plant with the hand-pushed planter described earlier and in which the
corn seed will sprout quickly.

On corn fields larger than about five acres, farmers who do not wish
to use herbicides will often begin their weeding regimen with what is
called a rotary hoe, a tool consisting of a series of wheeled, curved spike
blades. When pulled by a tractor at about 8 miles per hour, the rotary
hoes literally throw tiny germinating weeds out of the ground. The rotary hoe is run right over the top of the emerging corn rows, killing
weeds in the row as well as between the rows. Occasionally a corn plant
is hurled out of the ground, too, but not enough of them to be of any
consequence. As Dave says, "If the rotary hoe isn't yanking a corn plant
out once and a while, it probably isn't getting enough weeds either." A
rotary hoe can't be used in horse farming, because the horses can't pull it
fast enough. A spring-toothed harrow, with the teeth set in the most
slanted position, makes a fairly good substitute for horse farmers.

As soon as you can see the corn coming up, you can begin weeding
with wheelhoe and rotary tiller. On acreages too large for tiller management, cultivation with tractor or horsedrawn shovel cultivators begins
when the corn is about three inches tall. There is a wide assortment of
shovel cultivators available, old, new, horse, or tractor: one row, two row,
and so forth. When the corn is small, you will need shields for shovel
cultivators so that the shovel blades do not bury the little corn plants
with dirt. If you adjust the shields correctly, the shovels will roll (scrape)
just enough dirt in under them to bury tiny weeds in the row without
burying the corn. After the corn is taller, you remove the shields and let
the dirt roll unimpeded up next to the corn plants, burying the next
generation of weeds that tries to grow there.

Without shovel cultivators I use the rotary tiller, then straddle the
row and scuffle dirt with my feet into the rows to bury weeds. This is
not as slow or exhausting as it sounds, since the dirt is loose and mellow.
I move up and down the rows at a fairly good clip in complete control,
or at least my feet are. I can scuffle in just precisely the amount of dirt I
need to cover the weeds and not the corn. I get a chance to examine
every plant and every bug along the way, which I would not do if I were
seated on a tractor or horse-drawn cultivator. I can see where wireworms

and cutworms have struck, note with satisfaction the little black beetles eating smartweeds, and potato bugs eating horsenettle, and find, underneath a chunk of manure that did not get buried by the plow, two ground beetles of a kind that I know eat cutworms. Sometimes bluebirds, red-winged blackbirds, and grackles alight ahead of me, also intent on cutworms and other delicacies. Carol helps sometimes or our married children and their spouses, and our talk makes the work go pleasantly by. By July or even late June, the corn has grown tall enough to shade out germinating weeds, and cultivation is no longer necessary.

A new tool for tractors from 15 to 150 horsepower makes multi-row cultivation with rotary-tiller action (up to eight rows at a time) possible now. It's called the Multivator, from Mitchell Equipment in Marysville, Ohio.

Corn Harvest

Farm animals could actually harvest the corn themselves and often did in traditional farming. Lambs can be turned in the standing corn as early as August to graze the lower leaves of the tall corn stalks. They generally do not reach up and grab the ears. Next hogs can be turned in to eat the ears, and then the cows, horses, and sheep can overwinter on the stalks and the ears missed by the hogs.

But on most cottage farms, human harvest of the ears is necessary for feeding to chickens and livestock year-round, especially in winter, with a little corn saved for making cornbread and other delicious cornmeal pastries.

I start harvesting corn in late August, cutting green stalks that do not have nice ears and tossing them over the fence for the cows and sheep. This green corn is a good supplement to pastures that are often drying up in late summer. I have experimented with another trick I found in an old farm book, cutting the green stalk above the ear after the ear is well along toward maturity and feeding that to the livestock too. The ear matures, apparently no worse for having had its stalk decapitated. The practice may indeed benefit the ear because all the nutrients after decapitation are directed into the grain instead of the top stalk. My very tall open-pollinated corn is very good for this early harvest because the decapitation means the stalks will not be so prone to blowing over in a wind storm.

About September, after the kernels are dented and the milk line has receded halfway down the kernel to the cob, corn can be chopped by machine and put into silos. This requires a fairly large tractor, a chopper, a silage wagon, a blower, and a silo, so I do not believe that the process is right for the cottage farm of the future (too much expense). But some clever cottage farmer may figure out a cheaper way to do it, like running green stalks and ears through a brush chipper and using airtight plastic bags for storage. My feeding of green ears and stalks in August and September is my way of making silage. I let the animals' teeth do the chopping and their meat and milk do the storing.

Starting about September 25, maybe earlier, depending on the year, when the ears have bowed down on the stalk to point to the earth they know they must return to, they are dry enough to harvest and store. By then the husks are turning brown and the kernels are well-dented and beginning to "dry down." I do not bother to get the corn tested for moisture, because I know that in my slat-walled crib, the cleanly husked ears will continue to dry to a safe 14 percent moisture content without molding, and save me the natural gas drying bill that costs big cash grain farmers so much—13 to 18 cents a bushel and in wet autumn years more than that. In fact, drying corn artificially is one of the worst weaknesses of industrial corn production. The cost in money is horrendous in wet years and is a wasteful way to use one of our cleanest fuels, natural gas, because corn *could* be dried naturally in cribs in ear form like I do. The ears could then be shelled, if desired, for later marketing. But because corn farms are so huge now, and picker-sheller harvesters so labor-saving, it is fruitless to argue with large acreage farmers about how much money they could save by harvesting ear corn instead of shelled corn.

Ironically, ear corn brings a premium price over shelled corn at some elevators today because the cobs are in demand. Ground up cobs are used to polish metal. The exceedingly soft fluff on the outer surface of the cob is used in baby powders. Also the whole ear, both grain and cob, can be ground and fed advantageously to cows.

Some farmers of several hundred acres still do harvest ear corn and store it in cribs to be fed to livestock or sold later. One-row and two-row ear corn "pickers"—even new ones, but many old ones—are used for the harvesting. Old one-row pickers are affordable for the smallest farmers who do not want to hand harvest.

I harvest my acre or two by hand. With the pickup truck alongside, I go down the row, husking each ear and tossing it into the pickup, a few rows each evening and a few more on weekends. Maybe on a fine October weekday, I will just take the whole day off from this writing room prison and time warp myself back to 1940, when as a kid I followed my father around the field, helping him husk corn. Those were peaceful, gossamer days among the corn shocks, before the bombs dropped on Pearl Harbor—always the inevitable bombs—and ended for now or forever (who knows which?) the centuries-old stability of agrarian society. The war hastened the technological changes, cutting the past from the future as cleanly as a sharp axe cuts tree sprouts. The autumn of 1941 was the last time we hauled corn fodder to the sheep with horses and sled, with harness bells jingling.

As I walk down the corn row, I husk the ears in turn in one of two ways. Often the ear hangs down with the husk already loose enough that I can grab the end of the ear inside the husk with my right hand while my left hand grasps the stem end. A slight squeeze with my left hand while simultaneously a slight twist with my right, and the ear pops free from the husk and I toss it, all in one motion, into the pickup truck. If the husk is still tight on the ear, I slash it open with the husking peg tied to the fingers of my right hand, while at the same time, I strip the husk down the ear with my left hand. Then grabbing the ear with my right hand and snapping it off from the stem while holding the loosened husk in my left, the ear comes free. Although this is work, I view it as practice for the husking contests still held in our county. I can never beat the oldtimers who husked acres of corn like this when they were young, but I keep trying.

I try to keep all the corn silks out of the husked corn, and not leave any husk on the ears, since both the silks and husks would, in the crib, inhibit the circulation of air. But a stray husk or silk won't hurt. Nubbins are often still somewhat milky, which would also slow down drying in the crib, so I toss them to one corner of the truck and feed them right away to livestock.

I shovel the corn off the pickup into my little crib, built using a centuries-old design for drying the ears naturally. The walls are slatted, the slats (one-by-two-inch pine boards) are spaced about an inch apart for good air circulation. The width of the crib is about four feet, and should be no wider because that is about the limit that natural air circulation can penetrate into the ears in the crib (that is, two feet from either side). The length of the crib can be from here to kingdom come. The side walls slant out as they go up, so that rain, striking against the slats, will drip *down* and *out* of the crib instead of into it: very simple, but expressive of the collected genius of tradition. Small hatch doors along the top of one wall open, so I can toss shovels full of corn from the pickup into the crib. At the bottom of the back wall is another small door from which to shovel out the corn. I feed the corn to the chickens on the cob. They can peck the kernels off, saving me the job of shelling. Hogs can eat the corn off the cobs themselves too, and F.B. Morrison's *Feeds and Feeding* (Morrison Publishing Company, 1946) says they fatten just about as well on whole corn as on milled corn until they weigh 150 pounds, so that saves another shelling and grinding operation. When I do mill corn for the milk cow, or to finish out a hog or steer, or for little chicks that can't handle whole corn, I take it to the feed mill in town. I would like to have my own sheller and grinder, but at my level of production, it is cheaper to pay the mill to do this job.

Harvesting corn by hand is soothing work to me. I don't need to hurry. I have at least until Thanksgiving to finish, and could do ten acres, or even fifteen as our parents did, by husking a little every day all winter long. In fact I have mused about putting the whole place in corn, and husking just 25 bushels a day, hauling it to town, getting $50 a load, and being content with that much money and with the warm glow of Carol's eyes and the woodburning stove at day's end.

If I want to save the fodder (leaves and stalks) for livestock feed, I cut the stalks and make shocks of the corn like the Amish do. I used to shock some every year, mostly so the kids could pretend that the shocks were "teepees" just as I did as a boy. Dad didn't mind us playing inside our teepees because by making room for ourselves, we opened an ample space for good air circulation. I considered myself the Last of the Mohicans. Out in the field now, building a shock, I realize that as a farmer, I truly may be among the Last of the Mohicans.

I am not going to describe how to shock ripe corn for later feeding (see my earlier book, *Practical Skills*, for a complete description of that) because I don't think this traditional method is at all necessary on modern cottage farms. I have learned, quite by accident, that sheep, cows, and horses can be turned into fields of standing dead stalks after the ears have been harvested, and the animals will eat the leaves and top stalks on their own, thus saving all the work of making shocks. The animals in fact appear to relish this food even though the fodder seems too dead or dry to have much food value. The animals also knock down the stalk parts they don't eat, making a crisscross of plant residue on the soil surface that prevents soil erosion and provides cover for wild animals in winter. In the spring the rotted stalks are easily disked into the soil in preparation for planting oats. This "grazing" of corn stalks has become an important addition to my pasture rotation, since the stalks become available in late September and October right when I am often hurting for pasture. The fodder most relished by the animals is the dead sweet corn stalks and leaves, which they will eat right down to the ground. It would be quite practical to grow a field of sweet corn, sell the ears at farm market, and then "graze" the field with sheep, thereby profiting from two crops instead of one.

We harvest popcorn by snapping the ears off the stalks, husks and all. Then we strip back the husks but leave them on the ear. We tie the husks from three ears together and hang them on a wire in the garage, with metal can lids on either end of the wire so mice can't get to the corn. We do not shell the corn from the cob until we pop it, and in the unheated garage the corn on the cob seems to last indefinitely and still maintain good taste. We are presently popping corn that is three years old.

Some surplus sweet corn we use dried for parching. Throw a handful of the dry kernels into the popper after popping corn, to roast them.

Smoking meat with corncobs gives the meat a distinctive flavor. Corncob jelly is a traditional cornbelt delicacy, largely, I suspect, because it is made with a lot of sugar or honey. Cobs make a good bedding, under straw, for livestock. And if you want to apply for admittance to the ranks of real husbandmen, you need to be able to roll a cob across the barnyard under your boot ever so slowly and methodically when salespeople are trying to persuade you to buy something you don't need. If you proceed with the cob with great enough concentration, they will become impatient and, believing you are too dense to talk to, will leave.

CHAPTER 9

Cottage Mechanics

*A bulldozer in the hands of a wise man does good work; in
the hands of a fool even a spade is dangerous.*

The author's father in casual conversation, 1960

The boy pondered the task he had laid out for himself: to bolt
two hand hoes to the back of the goat cart his father had made for him
so that he could, by pushing the cart backwards, cultivate weeds between the rows of tobacco that seemed to stretch all the way across western Kentucky. He tinkered. He cogitated. He tinkered some more. No
one could have foreseen that fifty years later, in the 1990s, he would be
inspired by the contraption he was building to successfully design and
manufacture modern horsedrawn and human-powered farm machinery.
Nor could even the wisest prophet have predicted that between boyhood
and manufacturing, he would follow a successful career as a classical gui-
tarist and university music professor—and keep on farming, too. All he
cared about at that moment in the 1930s was how to get the tobacco
weeded faster and easier than with hand hoeing.

The cart cultivator, like most of Elmo Reed's ideas, worked. "Mo
and I could cover two acres a day pushing that thing," recalls J.B. Tyree,
a neighbor who delights in recalling the story. "That was the real beginning of both his horsedrawn Three-Point Hitchcart that is sold world-
wide today, and his hand-pushed wheel hoe cultivator for gardens." J.B.
pauses, reminded of another story that amuses him. "My granddaughter
is a track star. She took Mo's wheel hoe to the garden one day and *jogged*
up and down the rows with it. Mo said: 'Now that's real progress.'"

Elmo Reed would be the president and chief executive officer of the
contrary farmers if contrary farmers could endure such regimentation,
which, thank heavens, they can't. He exemplifies perfectly the combina-

tion of artistic and mechanical creativity without which farm work can become either a grueling exploitation of the human mind and body or, at the other extreme, a ruthless exploitation of nature. The same artistic creativity in Elmo Reed that flowered into good music, flowered into good farming. One of the travesties of modern life is that we have tried to separate "manual" work from "mental" work, as if that were possible. But Elmo says it so much better himself: "We have been taught that the world of art and poetry is far removed from the menial arts like farming. That's a myth and a very sad one because actually the so-called higher arts flow directly out of the so-called lower ones or they lose their vitality. If our rural society is lost, as now seems possible, what will become of urban culture?"

Much of the prejudice of "higher art" against mechanical artistry arises from a revulsion at the excesses of technology today. Traffic jams. Acres of barrels full of nuclear waste. A 200-horsepower tractor that replaces a hundred rural people and sends them to the cities to compete for jobs with a hundred people already there who can't find jobs. Gargantuan lumber machines that can swallow a tree whole at one end and spit two-by-fours out the other. It is almost amusing to listen to lumber companies hypocritically blame the spotted owl for lost jobs while their machines every year displace more thousands of jobs than a whole decade of environmental reform. If job loss in the timber industry is the issue, one could argue that the chain saw is far more to blame than all the endangered owls in the world.

It is precisely because of such issues that many people who claim to be environmentally conscious are suspicious of any machine larger or more complex than a pocketknife or a bicycle (even though they would never dream of giving up their automobiles or refrigerators). They harbor a notion that not only is playing a violin a far more noble pursuit than grease-monkeying an old tractor back into running order, but like Jean Jacques Rousseau, they believe that biology is something totally distinct from and culturally purer than mechanics. One of the blessings of farm life is that it teaches the error inherent in that point of view. Farmers see the utter dependence of biology and physics on each other. Birds use architectural construction skills to build their nests. Humans use the physics of air pressure to blow their noses. There is no instance of biological behavior that exists apart from mechanical principles. People who do not understand and are not willing to learn that, who because of this prejudice against manual arts even boast that they know nothing about how to build a house, repair a motor, sharpen an axe, or apply the physics of lever and gear to their work, will have a very difficult time as a farmer no matter how contrary they are. In my close observations, more cottage farms fail because of ignorance of mechanical arts than for any other reason.

I do not know how to articulate an accurate distinction between "bad" mechanical technology and "good" mechanical technology, or if there is one. If the use of lever and gear and heat and pressure and gravity, which is what all mechanical technology comes down to, makes work truly easier for the laborer without disrupting societal stability or ecological balance, then it is "good" I suppose. But almost always human behavior runs to excess and that is where the difficulty lies, not in the machine itself. You can kill a man with a pocketknife. But you can save a life with a monstrous 200-horsepower tractor that I consider an excess of technology, and in the 1978 blizzard when the roads were blocked by snow to all ordinary traffic, I saw that happen.

It seems to me that when artistic creativity becomes interested in agriculture—if we can increase the number of Elmo Reeds, in other words—the problem of "excessive" technology eventually takes care of itself. And I suggest this not only because excess is self-destructive in the long run anyway; when creative artistry embraces mechanical skills, the result invariably is a move *away* from excess. In agriculture, for example, the forerunner of the huge, four-wheel-drive, articulated steering tractors that make it possible for greedy farmers to hog more and more farmland, was built right here in Wyandot County by the Schmidt brothers, some of the most caring and gentle and mechanically creative men I know. They were not interested in hogging land but were simply enjoying the exhilaration that they felt from using their creative talents to build for a customer, Herb Walton, a tractor that would pull more than the World War Two tanks he was trying to farm with. And Herb was not a ground hog farmer either. He in fact was one of the first to pull back from the headlong race toward expansion and cash grain monofarming, and to begin trying to put into practice many of the tenets of organic farming. In the 1940s the need appeared to be for more powerful tractors, which is why he had the creativity to think of army-surplus war tanks.

And what do the creative wizards at Schmidt Machine Works specialize in today when the world is full of big farm machinery that no one can afford? They are front-runners in manufacturing what are called "after market" parts needed to repair the existing behemoths during these times when agriculture trembles before an unknown future. I have a notion that when the true costs of farming dictate a more efficient, small-scale approach to farm equipment, the artistry that built the Schmidt Machine Works and shops like it will lead the way back to sanity.

Lloyd Riggle, another locally famous machinist, is a consummate organic gardener. I doubt he or his wife goes to the grocery store once a month. They even make their own soap. They heat entirely with their own wood. But Lloyd is also a mechanical genius. He used to be a troubleshooter for International Harvester. Wherever a monster IH tractor or combine broke down in the U.S. and no one else could fix it, Lloyd was sent to the rescue.

What does he specialize in today? Restoring and repairing old *small* tractors! And we keep him fully employed. "It's more of a challenge to me to see if I can make a worn-out old tractor better than it was when it was new, than work on the big new stuff," he says. His specialty is installing alternators in place of old generators that barely kept batteries alive on older tractors, and rewiring a tractor's whole electrical system to fit the alternator. Lots of mechanics can change 6-volt systems to 12-volt systems this way, but when Lloyd does it, a tractor will start almost by snapping your fingers at it. With the motor reconditioned, such a tractor becomes not just an old restored tractor, but for all practical purposes a new, better tractor for modern small scale farming. What is needed, and what appears to be happening, is a Lloyd Riggle or two in every farming community.

Elmo Reed of course is the best example of what is happening. Can anyone imagine how students would greet a professor of agricultural engineering today if he told them that there was a good business opportunity in *horsedrawn* equipment? Only a very contrary and inventive person would divine such a possibility. "It didn't really take any genius," says Elmo. "I wanted to farm with horses myself when I retired from the university but it was hard to find horsedrawn machinery that wouldn't fall apart the second time across the field. Horse forecarts to which machinery could be hitched instead of hitching directly to the horses were already in vogue, greatly easing the work for the horses. All I did was put the standard three-point hitch on the forecart, along with hydraulics and PTO. That was still cheaper than a tractor, and I could then use all the modern machinery made for small tractors. In fact some of our customers pull my Three-Point Hitchcart with all-terrain vehicles (ATVs) instead of horses. That works fine and makes small scale farming more affordable to more people."

Today the small farm scene is full of PTO-equipped horsedrawn forecarts and there are half a dozen manufacturers of them. But Elmo, once the creative juices started flowing, went another step back into the future to make lower cost rather than higher cost new machines. He perfected *ground-driven* power-take-off shafts (PTOs activated by the forecart wheels as they roll over the ground), an idea Amish machinists had also been re-examining, and which was a commonplace technology in the old bull-wheel driven grain binders. Ground drive eliminated gasoline or diesel engines on the forecart, for those who preferred not to deal with piston power. But ground-driven PTOs which could turn fast enough to operate a rotary mower with the normal speed of a walking horse, required very sophisticated planetary gear designs. Elmo got the assistance he needed from high-tech engineers at Dana Corporation who had never thought to apply their technology to such an unusual purpose.

The Amish are past masters at using advanced technology to lower rather than to raise the cost of machinery, the opposite of what modern agribusiness does. Farmer Martin Schmucker near Fort Wayne, working with his neighbor, Melvin Lengacher, an Amish machinist, converted his baler and cornpicker to ground-driven PTO, dispensing with motors and using four horses to pull the machines. In operation, the balers and pickers are eerily quiet: the muffled rattle of chains, the whirr of gears, and the snuffling of horses are the only sounds. The two men even converted a heavy snap bean harvester to ground-drive.

The use of motors on forecarts to drive PTO systems is, however, more popular than ground driven systems (so far) with horse farmers and is an excellent idea for small farmers with very small tractors who do not wish to buy a larger tractor. The forecart engine finding favor with the Amish is a very new diesel made by Mitsubishi that uses a remarkably small amount of fuel—"*We've been running the cornpicker all day and burned only about a pint of fuel*," said the Amish farmer I visited recently, hardly believing it himself. "And the engine is so easy to start compared to the old Wisconsins we used to use." It also runs far quieter. The farmer showed me how any small PTO-powered piece of equipment could be hitched to the forecart's draw bar and its PTO shaft, just as you would do with a tractor. In front of the motor, handy to the farmer driving the horses, there is a little hydraulic pump, manually operated, to which hydraulic hoses from any piece of small machinery can be linked, again just as with a tractor. Working the handle on the pump backwards and forwards several times raises the implement, and then when the pump pressure is released, the implement lowers into operating position by gravity.

"English" farmers (which is, as I said, what the Amish call the rest of us) laugh at these arrangements, but it is the Amish who laugh all the way to the bank. The hydraulic pump costs a couple hundred dollars; the Mitsubishi engine, $3000. With the whole forecart assembly and a team of horses, the outlay of money is still less than half of what a new tractor powerful enough to do the same work would cost. Moreover, as my good Amish friend David Kline (he has a Ph.D in Contrary Farming from the University of Ornery Knowledge) patiently explains to visitors: "Our church allows tractors around the barn, but the power for field work must issue from horses. Otherwise, with tractors, the temptation to expand acreage would become too great and we would start competing each other out of business, as English farmers have done to their great detriment."

If you decide to farm with horses, subscribe to the *Draft Horse Journal* (Box 670, Waverly, Iowa 50677). That will give you all kinds of leads to horses and horse equipment. If you can visit Kidron, Ohio, in the heart of Amishland, for the big spring or fall farm machinery auction, you can get a crash course in horse and used small tractor machinery. Or visit the Pioneer Equipment Company at Dalton, Ohio, not far away, where modern horsedrawn machinery is still being made.

Most contrary cottage farmers at least for now are sticking with tractors for motive power, and in that regard they face a problem. New tractors, even smaller ones, are too expensive to justify for very small scale farms. Only from Russia and the Balkans are tractors being imported that are realistically priced for such farms, and dealerships for these tractors are few and far between. A few years ago, one of the U.S. farm machinery companies was thinking about manufacturing a very simple, long-lasting, low-horsepower, crude-looking tractor, easy to maintain and repair, but only for marketing in third world countries. I asked one of the executives why they didn't market such a tractor in America. "Oh, Americans wouldn't buy a tractor like that," he said. "They want soft contoured seats, cigarette lighters, twelve forward gears and two reverses, headlights, fenders, rubber tires, syncromesh transmission, shift-on-the-go, power steering, rollover bars, cab, gas gauge, front and rear PTO, dashboard lights, quick-mount hitches, gaily painted hoods, electronic fuel injection, four wheel drive, individual wheel brakes, radio, etc., etc., on their garden tractors." I replied: "The hell we do. We want what you intend to sell to Third World countries." Then the executive smiled and admitted, half in jest, what I think is wholly the truth: "Yeah, that's what we're afraid of. That would ruin our American market for fancy expensive tractors."

So until demand increases enough to lure farm machinery manufacturers back to making the tractors we really want (there is a company starting up to do that right now, in Ohio, but as of this writing it is not far enough along to want publicity), contrary farmers are buying Russian (Ozark organic farmer Eric Ardapple-Kinsberg says he has had no unusual problems with his Belarus) or are restoring and repairing old reliable American tractors. The market for them is being spurred by collectors and that is a good thing because the two groups—users and collectors—together amount to enough demand to generate a humming underground restoration supply business. (See the list of collector newsletters and magazines on page 184.)

There is in fact no better time than right now to equip the cottage farm with used machinery. Everything you need is available cheap because this equipment is obsolete on commercial farms which are locked into expansion and newer, bigger equipment. I saw a perfectly good four-row planter sell for $100 this spring at a farm sale. I recently bought a 50-horsepower tractor of 1967 vintage with a hydraulic loader on it, along with a 12-foot, hydraulically operated disk, plus a harrow, a cultipacker, and a three-point hitch plow all in good shape and all for $6000. By comparison a new "estate" tractor of only 30-horsepower (we call them "yuppy" tractors), which would be otherwise perfect for my size of operation, cost $18,000 or more with only a front-end loader. The smallest Belarus from Russia, at about 20 horsepower, costs around $8000.

To find good buys, you have to look hard at farm sales, study the classified ads of local papers and farm magazines, and haunt used equipment lots. But the real secret is to make friends with other farmers. Although I often criticize commercial farming, I don't blame most of the farmers caught in its ruthless economy. As things stand, they have little choice, and privately they grumble just like I do. They are, by and large, some of the most responsible and sensible people in society. My experience is that if you are friendly to them and acknowledge their expertise with a little humility even if you don't agree with their methods, they will be glad to help you, *once they realize you really are serious about farming*. Only with their help was I able to find the tractor and equipment I bought so reasonably. The advice one of them gave me bears repeating: "If at all possible, buy your equipment from other farmers. If you go through a dealer you will pay considerably more than you need to. Dealers are being squeezed just like farmers and selling new machinery is slow going these days. They have to try to make up the difference by inflating used tractor prices."

On the other hand, buying through an honest dealer generally means he will stand by what he sells. At auction or through private treaty, you run your own risk. If you are mechanically illiterate, buying a tractor from a reputable dealer may be wiser, even if it costs you a thousand dollars more.

On the other hand again, when you deal with farmers in your own neighborhood, where everyone knows everyone, they generally go out of their way to be honest about what they are selling. Everybody knows the few who lie. This is another advantage of a stable community. We don't need bureaucratic regulations to keep us from cheating each other.

Buying used farm equipment is fraught with peril for the amateur. Always check the oil in any tractor that you contemplate buying. If the oil is grayish instead of blackish, it has water in it, and invariably that means the block is cracked. But a cracked block is not necessarily the end of the world. I bought my first tractor with a cracked block (cheap) because I knew where there was a good block in a junkyard to replace it.

A piece of used equipment for sale might have missing parts. For example, the disk I bought cost me only $100 but I then had to figure in the cost of a hydraulic cylinder and hoses to raise and lower it, and these necessary accessories did not come with the disk. Chalk up another $70, at least. The plow I bought for $30 was made for a Category I three-point hitch. To make it fit the Category II hitch on my tractor required three bushings I could easily carry in one hand. Cost of the bushings? $38 at the dealership; $14 at one of the chain farm supply stores. An older, self-propelled combine in good shape that my nephew just bought for $1600, a real bargain, has belts on it that might need replacing soon. Just one of those belts costs $400. My 50-horsepower tractor may need overhauling in a year or two, which could mean spending over $1000. But still, compared to buying anything new, my nephew and I are way ahead. Even the smallest commercial farm combine made today is priced at nearly $100,000. A new 50-horsepower American tractor costs in the neighborhood of $35,000 to $40,000.

Most older motors were made to use leaded gas. When you burn unleaded gas in them, you should add lead substitutes or convert the valves for unleaded use. Most gas stations carry the lead substitutes. For a dollar and a stamped envelope you can get information on valve conversion and other matters from the International Society for Vehicle Preservation (Box 500046-1046, Tucson, Arizona 85703-1046).

The typical cottage farmer will usually have very little detailed knowledge about the tractor or other machine he buys, if purchased used. There may be problems with that particular brand and model that only a person who worked many years with it can tell you. In any event, there will be attachments, adjustments, and functions of the machine that you aren't familiar with. If you are new to farm equipment altogether, then your situation is very difficult indeed. You can't just "figure it out."

First you must get an operating/maintenance manual for your tractor or for other complicated machinery such as a harvester. For almost all equipment still in use, even if not manufactured anymore, these manuals are available through the dealers. Dealers will order the proper one from the company for you, if it is not in stock. The dealer will need the serial number of your tractor. In fact, whenever you order any part for anything, the parts service people can order accurately only if they have serial and/or model numbers.

Another source of written help are publications about old tractors put out by collectors. There is at least one for each major brand of "tired iron," as collectors call their tractors. Here is the list of magazines and newsletters that I mentioned earlier.

The 9N, 2N and 8N Newsletter (Fords)
Gerald Rinaldi, 154 Blackwood Lane, Stamford, CT 06903

Old Allis News (Allis Chalmers)
Nan Jones, 10925 Love Road, Bellevue, MI 49021

Green Magazine (John Deere)
Richard and Carol Hain, Rt. 1, Bee, NE 68314

Old Abe's News (Case)
Dave Erb, Rt 2, Box 2427, Vinton, OH 45686

Red Power (International, McCormick Deering)
Daryl Miller, Box 277, Battle Creek, IA 51005

Prairie Gold Rush (Minneapolis-Moline)
Roger Baumgartner, Rt. 1, Walnut, IL 61376

Wild Harvest (Massey Harris, Ferguson, Wallis)
Keith Oltrogge, 1010 S. Powell, Box 529, Denver, IA 50622

Antique Power (all tractors)
Patrick Ertel, P.O.Box 838, Yellow Springs, OH 45387

Oliver Collectors News (Oliver)
Dennis Gerszewski, Rt. 1, Manvel, ND 58256-0044

The second critical source of information is other farmers. There is nothing like mechanics to show the value of community knowledge and wisdom. How you cultivate these rich sources of information is more important than how you cultivate your crops. For example, yesterday our veterinarian, Dick, came to neuter our ram lambs and give all the lambs their over-eating shots. I had my John Deere loader tractor sitting in the driveway because we had just used it to carry the newly slaughtered steer up through the mud from the barn to where Neil, the butcher, could skin and clean it. (A sub plot to this tale of communal collaboration would star Neil as the hero who, more familiar with the JD 2020 and loader than I presently am, was able to maneuver it out of the mud hole where I got stuck and gave up.) Dick, the veterinarian, momentarily forgot about the lambs in his delight to find another cottage farmer who had purchased a JD similar to the one he had bought. He began to tell me things about my tractor I didn't know—really valuable information. He told me about a nylon universal joint on the oil pump that "almost always goes bad on older JDs," and how to replace it for a couple of dollars. "If it starts to rattle funny, replace it immediately because if you keep on operating with the worn one, you can find yourself facing a much more costly repair job," he said. He showed me what a peculiar mysterious slab of steel bolted to the draw bar was for (an adjustment for the three-point hitch). He pointed out my error in concluding that the plow I had bought would not fit the tractor properly. I just didn't understand how to make the proper adjustments. In fifteen minutes he gave me several years' worth of experienced knowledge that I desperately needed.

The most essential key to communal mechanical know-how for the cottage farmer is the country or village machine repair shop. Another sign of the deterioration of commercial farming is that farm machinery dealerships are getting farther and farther apart. That is, many of them are going out of business just like the farmers they once served. While the ones that remain still can provide excellent service via phone and United Parcel Service, when your tractor needs to be worked on, you can't ship it to that dealer by UPS. What is happening, at least in our area, is that the country repair shop is returning. There are two within driving distance of my tractor, one just down the road. Both are operated by an excellent machinist.

Dealing with tires is often the first repair or replacement job you have to do when buying used. Because *all* farmers, little, big, medium, conservative, liberal, contrary, or dittohead, need tires changed on the farm, the demand and therefore the competition has generated excellent service. New tires are available for every make and model of farm tractor still running, as far as I know. You will bless your tire serviceman many, many, many times.

In any event, don't view old tractors as you would old cars. Old tractors can be fixed indefinitely because most parts are still available. There's very little thin body metal to rust out.

Obviously, the drawback of buying used machinery is that you must expect to spend more time and money making repairs than with new equipment. But on very small farms, that often turns out not to be true after initial reconditioning. The reason is that on a small farm, especially one based on a pasture system, you will not be using your used machinery hard and so it may not break down as fast as new iron would on a commercial farm. This fact, along with the comparatively low cost of older used machinery, is leading to a somewhat amusing but advantageous situation. Many small farms, run by astute farmers with mechanical ability and an outside source of income, are much more over-equipped than the most advanced agribusiness farm. But for the little guy, that is not a detriment because it is not costly. For example, a neighbor who has farmed fifty acres all his life, and also worked as a machinist in town until retirement, owns four tractors that I know of, two trucks, and several cars, along with a very complete line of equipment. Most of these vehicles he literally rescued from the junkyard and restored. He has in them mostly labor, not much cash. Most of these tools are no longer depreciating, but are in fact gaining value from inflation and from collector interest and because, with only occasional light use, they are in excellent shape. By relieving his aching back and muscles, the many machines allow him to continue farming into older age. When he does sell the equipment at full retirement, it will prove to have been an excellent investment. Better than Social Insecurity.

In shopping for a used small tractor, be aware that there can be, between models of equal size, a price difference that has little to do with inherent value but a lot to do with popularity. For example, Fords from the late 1940s and early 1950s, especially the 8Ns and 9Ns, are favorites of the cottage farm set. So their prices are a little high for what you are actually getting in horsepower—they average about $2500 presently in north-central Ohio in good condition. A slightly newer and considerably more powerful Ford or Ferguson from the late 1950s and early 1960s will go for only a little more, $3000 to $3500 if you bargain, and give better service than the 8N or 9N; Fords are foolers in that throughout that period, the frame of the tractor stayed about the same size, but succeeding models packed more horsepower in the same space.

However, a less popular Allis Chalmers, a WD or WD 45, more powerful than an 8N or 9N but the same vintage, can be purchased in good shape for $1500. These Allis models, with both hand and foot clutch, represent about the earliest example of live power-take-off, a wonderful asset when using PTO-driven equipment. You can push in the hand clutch to stop the tractor's forward motion while the PTO shaft keeps on turning until you push in on the foot clutch. These old Allis models also have an accessory belt pulley that enables you to run a small stationary sawmill by belt, as well as many older stationary feed mills, threshers, corn shredders, and so forth. My Allis WD is not as handy to drive as a Ford 8N or 9N—the seat is too far away from the clutch—but in all other respects, I believe it is a better tractor.

Tractors and other machines offered for sale at antique shows are usually restored with love, not with the desire for money, and so are usually in good running order and worth more than what the restorer can get for them.

The Most Useful Equipment for a Cottage Farm

- After a *team of horses and forecart* or a *tractor*, the most essential equipment on a farm in my opinion is the *pickup truck*. If you own one as your primary or secondary means of transportation, you have already justified the cost and can use it almost free of charge for the hundreds of hauling tasks on the farm. If you can afford a four-wheel-drive model, you will find it a godsend in snow or mud country. But if you have a tractor, you can use it and your manure spreader (or a wagon or cart) in place of a four-wheel-drive truck for hauling through snow. If old pickups interest you, there's a publication that will help: *Plugs 'N Points* (Tom Brownell, Route 14, Box 468, Jonesboro, Tennessee 37659).

- The handiest tool on any farm is the *hydraulically-powered bucket* or *manure scoop* for farm tractors. There is a model to fit almost any tractor. A farmer usually gets a "loader," as it is referred to, to handle manure, but then finds scores of other tasks that loaders can do to save his or her back. Think of all the things you lift or drag in a year's time on the farm: sacks of feed; sacks of fertilizer; rocks; large dead animals; ricks of wood; logs; piles of dirt; gravel; bushels of apples, vegetables, grain; piles of compost as well as piles of manure; piles of leaves; corn shocks; piles or windrows or bales of hay and straw; fenceposts; rolls of fence wire; cement blocks; stacks of lumber. Not to forget: loading and unloading heavy machinery such as plows. Loaders can be used often as jacks: to lift barn beams so you can insert a new brace; to pull cornerposts and fenceposts out of the ground; to raise and hold stud walls or beams in place until they are properly nailed; to hold meat-animal carcasses up in the air while skinning and butchering them. You can use a loader as a little bulldozer to build a driveway and to scrape a gravel drive smooth, or to push dirt back into an excavated tile line. My father used his loader to scoop out an excavation for a farm pond, although I don't recommend that. A dam site can be too steep for a farm tractor loader, and heavy clay soil too much for such a loader to handle efficiently. The most bizarre use of a loader that I've ever heard of was lifting and freeing a ewe stuck in creek mud. I've had a ewe in this predicament and it took all my strength to free her. Next time comes the loader. But using tractors to pull animals out of swamps, quicksand, or mud is very tricky, since you could easily injure or kill the animal.

  Another important point: With hydraulic power on your tractor to run the loader, you are ready to use other equipment that can be raised with hydraulic cylinders connected by hoses to the hydraulic pump on the tractor. Adjusting the height or depth of plows, disks, and mowers with hydraulic power is heavenly compared to the older way where implements are raised or lowered by hand power or the motion of the implement's wheels. Hydraulically adjusted equipment is especially nice when you have small fields, like an acre or two, or for garden plots. You can raise or lower a disk precisely where you want to start cultivating and lift it up precisely where you want to stop cultivating. You can back into fence corners of the fields, lower the disk or plow, and cultivate the entire space. These maneuvers are very difficult with pull-type implements.

The above use of hydraulic power also makes possible the *three-point hitch*, the second (some would say the first) most essential tool for your tractor or forecart. Almost all smaller cultivating tools—mowers, scrapers, pole lifts, fork lifts, box lifts, sweepers, planting and harvesting tools, post hole diggers, and winches—are now made to fit the three-point hitch system, allowing the operator to raise and lower any such equipment hydraulically.

- Primary cultivating tools you will need include the *plow, disk, spike tooth harrow, spring tooth harrow*, and perhaps a *cultipacker*. My argument for the plow over the chisel plow is made in another chapter. The disk and spike tooth harrow are, in my experience, the most widely applicable (and lowest-cost) tools for working plowed ground into a seedbed. If there is enough plant residue from last year's crop on the soil surface to plug up a spike tooth harrow, the spring tooth harrow can sometimes replace it. If there is a whole bunch of residue, you will have to work the ground with the disk alone and attach some kind of drag behind to level the disk marks. A length of log will work as a drag. I once talked to a farmer who used a small cedar tree for a drag—these trees grew in abundance on his farm. Disk and harrow can also be used to work up soil without plowing it first, especially if the land is somewhat bare, as after corn.

- A *cultipacker* would not be considered an essential tool on most farms, and I have only lately acquired one. I now believe that a packer is crucial to cottage farming if you are emphasizing grassland crops rather than grain crops because you will sometimes be planting grass and clover in summer and early fall when the soil is usually dry. If you cultipack the dry seedbed after broadcasting the seed, you will enhance your germination rate tremendously over broadcasting without cultipacking. Although a regular grassland drill equipped with cultpackers is expensive, a broadcast seeder and an old cultipacker will cost you peanuts.

- If you are as anxious as I am to avoid pesticides whenever possible, you will need a *weed cultivator* for your row crops. I've used a garden tiller for years to cultivate between corn rows, but am now increasing the amount of corn I grow a little, and plan to get a tractor cultivator when an affordable one comes my way. If you have a three-point hitch on your tractor or forecart, you can easily find two- or four-row cultivators, new or used, that are relatively inexpensive.

  If no-till chemical salespeople give you a hard time about how row cultivators increase erosion, just say "Monroe J. Miller" to them. Monroe is an Amish farmer who has for years been horse-plowing and row-cultivating hills too steep for tractors to maneuver safely on. He does it with great intelligence using the art of contour strip farming, alternating narrow strips of clover with narrow strips of corn laid out perpendicularly to the slope of the hills. I have stood on those hillsides and been astonished to silence. There is less erosion on his hills than in "no-till" fields where herbicides are used to kill plant growth and then crops are planted down through the residue of the killed plants without disturbing the soil. When I saw him at a horse-plowing contest this spring, he told me that early on, when he was disputing the claimed advantages of chemical-soak no-till farming over his method, the no-till champions from Ohio State University took him to task for plowing and row-cultivating steep hills. The county agent came to Monroe's defense and pointed out that observations and soil samples taken from Monroe's steep, contoured hills indicated very little, if any, erosion. Also no compaction, which would be rarely true in no-till farming, and a very high organic matter content. Of course everyone who knows Monroe Miller knows that would be the case or he wouldn't farm those hills.

- You will need a *broadcast seeder*. You can get large PTO-driven ones for the tractor, but the little, over-the-shoulder model is big enough for the small farm.

- Unless you are young, strong, very talented with a scythe and have less than an acre of hay, you will need a *sickle bar mower* for cutting hay and a *rotary mower* for clipping meadows. You could get by with only a sickle bar mower if you keep the blade sharp. With some of the new rotaries designed for hay, like the "Worksaver Haymaker" (Litchfield, Illinois 62056) you could possibly get by with only a rotary mower, but I think it wise to have both if you can afford to. Both are available new or used to fit three-point hitch systems. Flail mowers and the new drum disk mowers are wonderful but more expensive than a cottage farm can normally justify.

  Not so incidentally, for understanding and maintaining sickle bar mowers and the whole range of older farm machinery, John Deere put out a book for many years called *The Operation, Care, and Repair of Farm Machinery*. I have the fifteenth edition and the twenty-fifth edition and guess that they were published about 1937 and 1947. Copies of this book constantly pop up in used book stores and antique shops. The machinery described is all John Deere, but the discussions and drawings are detailed enough that you can apply the information to other makes and models. I always read through this manual enviously because the horsedrawn and tractor-drawn machinery discussed in these editions would be perfectly apt for the cottage farm today.

- You will need a *hay rake*. Older ones regularly sell at farm sales for under $50.

- You can harvest hay loose the low-cost way as I describe in chapter 10 for small acreages, or, for over about five acres of hay, get a *baler*. If I used a baler, I would prefer one of the models that make small rectangular bales, not the ones that roll up big round bales. The latter cannot be moved by hand and are usually stored outside, which means the outer layer of hay deteriorates. Steve Gamby, a neighboring dairy farmer, tells me that he thinks the small bales are better for small operations and generally provide a higher-quality hay.

- A *manure spreader* is almost a necessity. Used ones are available generally at a big advantage in price over new ones.

- I discuss *grain harvesters* in chapter 10. You can hand harvest the corn from a small acreage, but any other kind of grain in fields larger than a garden plot will require a mechanical harvester. Consider first having a custom harvester do the job, meaning a farmer in the neighborhood with the right equipment. A second alternative would be to buy a used *combine* as I have done. Ten- to thirteen-footers (the width of the swath they cut) in the self-propelled category are obsolete in commercial farming, used now only for combine demolition derbies. I often muse on that tragi-comedy. Most of those combines were never paid for: bought with borrowed money and traded in on bigger ones with more borrowed money. Now they are "worthless," the debt hanging over them to be paid by a future generation. No wonder farm boys in demolition derbies crash them to smithereens with such glee. I think deep in their souls they know they've been had.

  An old obsolete combine still in fair shape can be purchased for about $2000 if you wait for the proper chance. However, unless you are as lucky as my nephew, most of these combines are about worn out and will need considerable repair. Two years ago at the Gathering of the Orange, the annual antique Allis Chalmers fair, there were two pull-type, restored AC All-Crop combines for sale at under $1000. Since new parts are still available for the All-Crop, these combines are good buys. Mine, which I purchased for $50, has served me in good stead for twelve years. It rolled off the assembly line in 1948.

- *Fencing tools* are a necessity for all farms with livestock, especially where intensive grazing is practiced. If you prefer woven wire fences for your boundaries, you will need one or a pair of traditional fence stretchers, available new by mail from farm supply centers like Nasco Hardware (901 Janesville Avenue, Fort Atkinson, Wisconsin 53538). Stretching with a tractor is not very satisfactory and can be dangerous. Great force is required to stretch a woven wire fence properly. I can tell you how to do it in writing, as I have done in my out-of-print book, *Practical Skills*, but you will have to learn how to do it, hands on, from someone in the flesh. The critical art is all in setting the corner posts solidly enough, not in operating the stretchers. Stretch only in a straight line. Do not try to go around curves. Always set a brace post with every end post, with a brace bar between them and a guy wire to tighten from the bottom of the cornerpost to the top of the brace post. There are other ways to make a corner post solid with bracing, but this double post arrangement, the posts about five feet apart, is easiest. End post and brace post should be sunk into the ground three and a half to four feet, with at least five feet of post above ground.

Woven wire fencing comes in all kinds of heights and gauges (wire diameters). The top and bottom wire of your fencing should be the thicker 9 gauge diameter. The other wires should be no thinner than 11 gauge (the higher the number, the thinner the gauge). If you buy cheaper, thinner gauge you are throwing your money away. The width between the stay wires (the vertical wires) can be four inches, six inches, or eight inches. If you use the better (and more expensive) four-inch stay size, get it with the graduated spacing between the longitudinal wires, narrower at the bottom so little pigs and lambs can't get through, and so ewes can't get their heads caught in it. The cheapest worthwhile fencing I have found measures forty-two inches high with an eight-inch stay, top and bottom wires 9 gauge, the other wires 11 gauge. Sometimes at the farm supply store, I have to root through the rolls of fencing to find this combination. Do not expect help from salespeople, because no one knows anything about woven wire fencing anymore.

The higher the copper content in the wire, the longer it will last and the more expensive it will be. Most of the chain farm supply stores (like TSC and Quality Farm and Fleet here in northern Ohio) don't even carry the high-copper-content fence (like the better kinds of Red Brand) but if you can get it reasonably, the higher price pays. Why put up cheap fence considering the work you put into it?

At almost every farm sale an old set of fence stretching tools will sell cheaply because livestock farmers are turning mostly to various types of electrified fence that require no heavy power stretching. Some farmers prefer to put up three to five strands of barbed wire, stretching each strand individually. This can be done with a block and tackle or the singlestrand "one-man" stretcher, available from mail order farm supply houses. I know that Lehman Hardware (P.O. Box 4779, Kidron, Ohio 44636) carries the one-man stretcher, and this store sells nationwide by mail. Every cottager should have Lehman's unique "non-electric" catalog anyway. A *steel fence post driver* is a necessity, and a *fence post digger*, too, for wood posts. You can rent power hole diggers.

I use a hand-powered *post hole digger*. I build about two hundred feet of fence a year on the average, and don't need any more speed than muscle power. Muscles are so much quieter than gas - gulpers.

The topic of fencing always gets me embroiled in argument with other livestock farmers. Many say New Zealand-type electric fence is cheaper and easier to put up than traditional woven wire. Being contrary, I of course disagree. Traditional woven wire fences, built right, last thirty years, and I don't know of any New Zealand-type fence that has yet been up nearly that long in this climate, so how does anyone know it's better? And by the time you buy all the doo-dad connections, insulators, and individual wire tighteners you need for New Zealand type fence, I'm not sure it's that much cheaper or faster to erect either. The only advantage to New Zealand fencing is that it is readily available in all kinds of varieties, and with ample instructions on how to erect it.

But assuming these new electrified fences are cheaper than new woven wire fencing, my comeback is that I don't use new posts and wire when I build a traditional woven wire fence. Thanks to the great Interstate Highway System, there is always someplace within hauling distance where the fencing along the highways is being replaced—miles and miles of it. A wonderful opportunity, because most of this fencing still has fifteen to twenty years of life in it, maybe more, and it is far better fencing than any new stuff farmers can afford. It is made of all 9 gauge wire and corrosion resistant steel. The steel posts the state uses (good old tax money) are also far superior to any you can buy on a mere taxpayer budget. The first load of this super-duper fencing I got for nothing, but the companies tearing down the wire quickly caught on so now they charge a little for it. Nevertheless it is still a great bargain.

For end posts and wooden line posts I have secured permission to chain saw discarded electric-line posts at the local utility station into proper lengths (seven to eight feet) and then split them into two, four, six, or eight posts, depending on their size. Working leisurely, I can make twenty-four posts in an afternoon that are of far better quality than the wooden fence posts you buy for $3 to $5 each. Unsplit, the butts of these poles make terrific end and corner posts, much more durable than the $12 to $15 end posts sold for this purpose. Best wages I make all year.

The upshot is that I have fencing which, now that I have taught myself the fine points of fencebuilding, will stay horse high, bull strong, and hog tight for a quarter century for very little money. And I don't have to worry about the electric fence shorting out and the livestock stampeding away with nothing to stop them between here and Chicago except the fencing along the superhighways.

- If you cut wood for fuel, you need a *chain saw*, an *axe*, a *peavey*, two *wedges*, and a *steel splitting maul*. A gas-gulping splitter is nice and noisy but not necessary. A splitter that runs off the hydraulic power of your tractor is much better, but also not necessary until you get too old to swing a splitting maul. Use the wood you can't split by hand for fireplace back logs or for boiling off maple syrup and lard in open kettles.

- If your farm is in an area like ours where the clay soils drain poorly, and you absolutely need tile drainage at least in the wetter places, you might find that an investment in *tile trenching spades* is a good idea. Hand-digging tile ditches is hard work, and if you have more tile to put in than, say, two hundred feet a year, you will probably want to get a knowledgeable drainage contractor with a power ditcher to do the work. A power ditcher digs deeper and therefore does a better job of preparing for drainage than you can do by hand. But I learned this spring that digging three hundred feet of tile trench about two feet deep was not that physically difficult after all. If at sixty-one I can dig that much in my spare time in April, younger people ought to be able to do six hundred. It's kind of pleasant work in a way, what I call pure relief work: Mostly muscle, very little brain.

- I use a *shovel* or *spade* to cut weeds out of the pasture. Easier, it seems to me, than using the more traditional hoe for this job. With one thrust you can sever the tap root of a bull thistle, mullein, sourdock, or burdock. Swinging with the hoe usually takes at least two whacks. But my cousin Raymond in his old age devised a neater way to cut pasture weeds with the hoe. He sawed the handle off his hoe half way down, mounted his trusty riding lawnmower, and set sail across the meadow in high gear in pursuit of weeds. As he passed them, he would swing his short hoe viciously at the base of the weed, decapitating it without stopping. He reminded me of an old polo player not quite ready to give up the game.

- Time was when I would have put the *rotary tiller* at the top of the list of cottage farm tools. I am not so sure anymore because rotary tillers stir up the soil so much, like an eggbeater. Still, the rotary tiller is the best tool for small plots unless you prefer spading and working the ground with hand tools. Hand tools are better for small, raised beds. Rotary tillers are excellent for light weed cultivation and for breaking up clods.

- *Pitchforks* are still needed on cottage farms and not just for running off unsolicited salespeople. I use a four-tined manure fork even for hay, although for hay a regular, three-tined hay fork is better. It is lighter and slides into and out of the hay easier. I also have a large, four-pronged fodder fork that is good for handling leaves, cornstalk shreddings, and grass hay cut with the rotary mower. I also find frequent use for a large fork of heavy tines positioned very close together. I call it a cob fork but have heard it referred to as a stone fork and as an ear corn fork. I use it to clean up the manure that is too fine for the regular manure fork. It works especially well to handle composted chicken manure or any other compost.

- If you raise grains, you will want a *grain scoop shovel* and probably a *bushel basket* or two. We buy rubber buckets occasionally to carry feed to the livestock and chickens, and latch onto plastic buckets originally used for other products. I am always on the lookout for free steel barrels for water or grain storage. As a general rule, cottage farmers will beg, borrow, or steal any bucket-sized container that comes their way. There is always a use for these and they are continually rusting through or cracking.

- You will need a *lawnmower* unless you can maneuver in your yard with your farm tractor and field rotary mower. The smaller the lawn, the better, I say, but we have a huge one at present. I intend to put a fence around most of it and graze sheep there. (Carol doesn't know that yet.) A small lawn can be mown with a push mower. There is nothing more ludicrous in the modern world than the person who rides an $8000 power mower around a postage stamp lawn on Saturday morning and then tries to jog off the fat on Saturday afternoon.

- Buildings are tools. I think a *heated repair shop* is among the best tools a farm can be equipped with, because then you can use winter hours comfortably and even pleasantly for the unending and immensely money-saving tasks of machinery repair.

    I've described our *corn crib* and *grain bins*, *hog* and *chicken sheds*. If you intend to raise animals, buying a homestead with an old *barn* on it saves yourself a lot of money. Buying lumber today is almost as expensive as buying silverware. Even broken-down barns contain a lot of good wood that can be reused. Our barn has parts of two older barns in it. A neighbor did one better: purchased a building elsewhere, sawed it in two, moved the sections to his barnyard with tractor and trailer, and put it back together again.

    Moving buildings is an old tradition around here, for in truth, lumber has always been high-priced for thrifty people. Two homes within a mile of us were moved in, and another was turned around, heaven only knows why. Two miles south, there's a machine shed in a barnyard that was formerly a church in a nearby village. The village has long since disappeared. On the next farmstead east, the barn is constructed of lumber that came from our first courthouse, torn down in 1900. Rural communities were recycling before the word was invented because they couldn't afford not to. We have a waste management problem today simply because there is too much money for new products in circulation.

How you arrange your barn depends on what you are going to raise in it, of course. Looking back now in hindsight, my advice is to make very few partitions, hay mangers, and pens until you actually start keeping animals in the barn and your needs become apparent. Best to have two floors, a hay loft above, with holes through which you can shove hay to the mangers below and end doors through which to load hay into the loft. Go look at traditional barns for ideas. Older agricultural books and magazines, especially those published before 1920, sometimes contain great drawings of barn layouts and accessories. The most instructive lesson I've received in this regard is to have witnessed an Amish barn raising. I truly doubt whether the most automated chicken factory in Arkansas is half as efficient as the way the traditional Amish barn is laid out for the care and feeding of farm animals by hand labor.

- By all means keep a pair of *pliers* and a *pocketknife* in your pocket at all times, and a roll of *baling wire* close by. A beak-nosed fencing pliers is a blessing when repairing fence. You would be wise to buy seventeen hammers and thirty-three screwdrivers to strew around your property so that you can always find one when you need it. I often carry a magnifying glass in a sheath my daughter made for me. The original reason for the glass was to observe bugs up close as I worked in the garden or walked the fields. Lately though, I find the magnification a great aid when working on machinery. It's hell to get old.

- Last but not least, keep a *grease gun* and an *oil can* handy. I am forever amazed by how often a malfunctioning tool or machine needs only lubrication to make it work right again. The worn bearings of old equipment especially need to be kept bathed in grease. There must be sixty grease zircs on my old grain combine (I've never counted them), and the fact that every one of them has received three or four pumps of grease from the grease gun before each day of use is the main reason that machine is still running half a century after it was manufactured. Years of life can be added to the steel parts of shovels, spades, cutter bars, disk blades, and plow moldboards if they are swabbed with waste oil between uses.

In general, learn as much as you can about hand tools because if you get exactly the right one for the right job, the work is so much easier.

We are losing the lore and knowledge of proper hand tools—an axe handle is shaped the way it is for precise reasons that few can appreciate unless you try to swing an axe not shaped the traditional way. The spade I bought for digging tile trenches is a new thing, poorly designed for the job. I tried to use it until my back about gave out and my foot got sore trying to push the blade into the soil. Only when I tried cousin David's hefty old spade, made specifically for digging tile ditches, did I understand why this menial work need not be repugnant. Even if you don't really know the fine points of hand tools, if you buy used ones at farm sales, you are much more likely to come away with something good, and at a fraction of the price of nice, new, poorly designed junk.

Keeping tools properly sharp is another skill whose importance can hardly be over-emphasized. My friend Roy Harbour's memory of farming fifty years ago is sharper than his hoe. He says his father always honed a hoe blade on both sides of the edge, unlike the way hoes come from the store. "You never had to swing my father's hoes," says Roy. "Just nudge the weeds with the blade and they turned up their toes." Once Roy gets started, he can't be stopped easily. "Dad would heat an axe blade a little, and then pound the blade out pretty flat to back about an inch from the edge. Then he'd spend a whole evening sharpening that thing till it would cut hair. I am not kidding. With his axe and a little practice you could cut a four-inch sapling off with one whack."

No wonder it took "civilization" less than a century to turn millions of acres of primeval forest into a great big ball diamond.

CHAPTER 10

Winter Wheat, Spring Oats, Summer Clover, Fall Pasture

The people are the most important element in a nation; the
spirits of the land and the grain are next; the sovereign is
the lightest.

Mencius, circa 300 BC

Grain farmers and fishermen have no trouble understanding each other. One plows the land; one plows the water. Both must bow to the fickle antics of nature in their search for food. Their inability to control the weather breeds in them a patient pessimism that endures century after century, outlasting all of the power hungry zealots and bumbling bureaucrats. Compared to nature, zealots and bureaucrats are a piece of cake.

The similarity between grain farmer and fisherman extends even to a visual likeness between grain field and body of water. Growing grain takes the moods of the wind as water does. The Polynesians distinguish dozens of different plays of wind on water and these seafaring people have names for each of them. I yearn for special words as well for the winds that move the amber waves of grain. But I must make do. Catspaws ruffle the surface of the growing wheat and oats as on the surface of the pond. Whitecaps of grainheads nod in the breeze like the undulating swells of a seascape. My eye can follow a whirlygust across the field as the grain stalks bend before the wind in swirls made visible by a contrast between the lighter-colored stems and undersides of the plants and the darker tops: whirls of color from ivory to jade when the wheat is still green; of burnt siena to gold when the grain is ripe. The understory clover thickens to a luxurious green as the grain matures, and then a fitful, fretting wind parting the taller wheat gives the impression of emerald raindrops dimpling a golden lake. I think of a painting by Manet.

At dusk comes the most bewitching sight of all. The sun, already below the horizon, still casts enough red into the skies that the color reflects horizontally on the ripening grain and the wheat heads look like tiny embers in a dying fire. As the embers fade, fireflies come up out of the grain to mimic that effect, the females on the wheat stems, the males in the black air above the field: hundreds of thousands of them blinking love messages in an utter serenity of silence.

I have seen grain and water unified after a June downpour sent the creek cascading out of its banks. The flood waters rose in the wheat field until only the ripening wheat heads stuck above the surface: a yellow foam on the water lapping at the very brim of a red-winged blackbird's nest built deftly into the tops of a bunch of stalks, and looking like a tiny, moored boat. Mother blackbird perched on the edge of the nest, singing as if she had not a care in the world. That night a chorus of frogs made merry as the water receded. Two weeks later, I harvested the grain in a billow of dust and wondered, from the tractor seat, if I had only dreamed of the frogs.

The beauty of soft red winter wheat (that's what we grow in Ohio) is not just a harvest-time show, but an all year event. After I finish the wheat harvest and gather the straw for winter bedding, the interplanted clover grows again with great vitality, unhindered now by the shade of the wheat, and intent upon blooming and setting seed before winter. By September the field that was green in June and golden in July is pink with clover blossoms over which dance legions of multi-colored butterflies and bees.

In a nearby field, where I harvested oats in July, I plant the next year's crop of wheat in the last week of September, after danger of Hessian fly infestation has passed. By the end of October, the field resembles a lawn in June. With a border of woodland in fall colors, the green wheatlands become my delight, especially as they grow still greener into somber November.

The wheat can also double as sheep or even cattle pasture in the autumn without harm for the next year's harvest, if the grazing is not severe and the animals are not allowed on the field when the ground is muddy. Sheep are better for this. Their dainty feet do not trample the wheat as much.

And what greens up first in the drab days of March? The wheat fields of course. It is also in the wheat that I perform my first and favorite planting rite of the year: broadcasting clover seed while the wheat is still dormant or even after the little blades begin to grow again. In the Ozarks, where spring plantings are more apt to be accompanied by warm weather, folklorist Vance Randolph recorded in the 1930s that menstruating women, working naked by the light of the moon, were thought to give the crops special advantage. Success would be insured if sexual intercourse were performed on the seedbed. I asked a modern Ozark farmer if those superstitions continued today. "That's not superstition," he replied with a chuckle. "That's just a good excuse."

The Culture of Wheat

Winter wheat has immensely practical advantages for the cottage farm. It is the only crop planted in the fall in the north (unless you count the intriguing possibility of growing turnips and kale for winter pasture), and so it spreads out the work load for both farmer and machine. If I were to plant no wheat, I would have to plant more oats in the spring, which would possibly mean larger equipment and certainly more labor right when I should be getting ready to plant corn. Also fall wheat planting occurs after oat harvest and before corn harvest, avoiding the scheduling conflict that results if soybeans are included in the rotation (soybean harvest invariably occurs right when you should be planting wheat). Moreover, fall planting enables the wheat to get established without weed competition, since weeds grow only weakly, if at all, in fall-cultivated soil. If the wheat follows oats in the traditional manner, the field can be disked several times while it lies fallow between oat harvest and wheat planting, a great advantage when a field has become infested with weeds from poor chemical farming or a lack of proper rotations.

The interseeding of clover (and often timothy along with it) in wheat makes it unnecessary for the farmer to spend time and money on seedbed preparation exclusively for the clover and timothy, since these crops germinate well enough on the wheat's seedbed. The timothy seed is so fine it does not need to be covered with soil. Just broadcast it right on top of the ground immediately after the wheat is sown. It comes up with the wheat, but its blades are so fine they are barely noticeable until the following year. The clover is added in early spring when the soil is still relatively bare. Rain, or recurring freeze and thaw, will allow the tiny clover seeds to sink far enough into the soil for good germination, which is, after all, how nature plants seeds.

I have sown wheat and oats for twenty years by walking across the field, back and forth, cranking a little broadcast seeder slung over my shoulder. In one pass across the field, the broadcaster sows a strip about thirty feed wide, which is about what a $15,000 drill will cover in one pass. But the broadcaster sells for less than a hundred dollars. Although I now plant only a couple of acres this way, I have sown ten to twenty acres over the course of two days in younger years and not expended any more energy than I would have in jogging a couple of miles each day. My friend and neighbor Dave used to seed seventy acres of clover in a season this way.

Plant two bushels of wheat per acre when broadcasting, and eight pounds of red clover seed. The rate for timothy is about two pounds per acre. There is a gauge on the broadcaster to set the proper amount, but I do not pay much attention to it since the actual seeding rate will depend on how fast I walk and crank the whirling propeller that scatters the seed evenly over the ground. If I can see two wheat seeds per square inch on the ground, I figure I am sowing enough. I prefer to err on the side of too little rather than too much with grain, because the clover will do better in a thinner stand of wheat than a thicker, and I value the clover more than the grain. I run a check on myself as I plant by noting whether or not I am getting a half bushel spread with the wheat for every quarter acre. I can decrease or increase the rate by either opening the aperture out of which the grain falls onto the propeller or by walking slower while I crank. With the red clover seed, I make sure every quarter acre gets at least two to three pounds.

Before seeding I disk the old oat stubble several times in late August and early September to clean up weeds and incorporate the straw and stubble remaining after oat harvest. After planting I go over the field again with the spike tooth harrow set as lightly as possible, to further level the soil surface and cover the seed a little.

A surer way to get a good stand is to plant the wheat with a grain drill, about an inch deep, maybe two inches in dry falls. Old drills are common at farm sales and inexpensive, but my grain growing has not suffered for lack of one, so I continue to do without. In dry weather, going over the field after seeding with a cultipacker to firm the soil is a good practice.

Between planting and harvesting wheat and its companions, timothy and clover or alfalfa, there is nothing else to do except admire the grain if it promises a good crop, or grumble about it if not. Weeds can become a problem where the traditional rotations are not followed or where the farmer relies totally on herbicides for weed control. One of the ironies of modern farming is that now with all the streamlined herbicides, Canada thistle and giant ragweed (to name but two) are a far worse problem than they were *before* herbicides.

Once I have my little fields into proper rotations, weeds are generally not a problem. I watch for stray thistles and sourdock, and cut them back before they bloom. A sourdock here and there that is not allowed to produce seed can actually help the soil by punching its long carrot-like root down into subsoil, increasing soil porosity and pulling up minerals to the soil surface just as legumes do. After harvest, a mowing to clip the wheat stubble sets weeds back again, and then the clover comes back strongly enough to choke them out the rest of the year. Mowing the clover for hay twice or thrice the next year or two, along with some pasturing the second year, keeps the weeds at bay. Then the land is plowed for corn and the corn cultivated intensely to further discourage the weeds. Oats, following corn in the rotation, has its own allelopathic (herbicidal) effect on grassy weeds and provides a measure of control until after harvest. Summer fallow cultivations after oat harvest, or after mowing of clover if clover was interseeded with oats, bring the rotational weed control practices back full circle.

Eventually the good farmer—following a rotation of corn, oats, wheat, hay, and pasture and then back to corn—will keep his fields relatively weed-free without herbicides. If a special weed problem does occur, then a small spot-spraying of herbicide to the problem area can be very effective, but only under the mantle of proper rotations and cultivation. I am sorry to have to disagree with the organic purists who liken herbicide applications to human drug use and say that it only takes a teaspoonful to start an addiction. A grain field is not a mentally sick human being. On the other hand, depending on chemicals alone, with heavy, routine sprayings, only insures that you will have to spend lots of money on a large amount of chemicals again the next year. Every year the cost of chemical farming creeps up. The standard single application of herbicide to soybeans is now up to $30 per acre, more than twice the cost of cultivation. Often more than one application of herbicides becomes necessary, and in no-till farming, applications of insecticides, which are more toxic than herbicides, become necessary as well. That money allows the herbicide companies enough profit so that they can afford to spend huge sums of money on those sickeningly hypocritical advertisements that convey the idea that using herbicides is the way to preserve the family farm. Herbicides are the single most useful technology by which mega-farms can keep on expanding acreage and thus force family farmers off the land.

The first reward of the grain is in eating it yourself. The smell of nirvana is homemade bread baking. The latest food guides advise a diet about half of grains, and about a fourth each of fruits/vegetables and meat, but sometimes I think I could almost live on home-baked bread (spread thickly with butter) and homemade pasta.

Soft red winter wheat is used mainly for pastry flour. White wheat (durum) is generally used for pastas. Hard red winter and spring red generally are milled for bread flour. We use our wheat for all purposes except pasta. We mix a little white bread flour with our own home-ground whole wheat to make the bread lighter. Although pasta is generally made from durum wheat flour, we use regular unbleached flour in a recipe of one pound flour, five eggs, and one half-teaspoon of salt. John Rossi, of Rossi Pasta, now a very successful gourmet pasta business, told me years ago not to use the usual pasta flour called semolina because it was a bit too coarse for the little home-style, hand-cranked pasta machines like ours. He recommended the Sapphire brand of durum. We could probably grow our own durum but that is just one more of the ten trillion projects I have not found time to explore yet, and regular flour works fine for us.

At any rate, homemade pasta is even more of a treat than homemade bread. It is lighter and tastier than what you normally buy at the grocery. Most commercial pasta (*not* Rossi Pasta) is manufactured by extrusion—that is, the moist dough is squeezed by pressure through a screen, which heats and half-cooks the pasta and makes it heavier and pastier.

We feed our own wheat whole to our chickens. I also grind some in with the corn and oats for all the animals. In feeding wheat whole to hogs, it should first be soaked for a few days. If a pig eats a lot of dry wheat, the grain can swell in its stomach and kill the poor thing. Soaking on the other hand begins the fermenting process, which could lead to slightly inebriated but happy hogs. My father soaked even milled grain for his hogs, and the joyous sounds of the pigs at feeding time brought to mind, well, the noise emanating from a downtown Chicago singles bar during happy hour.

My father-in-law used to tell a story I can repeat now because all the people involved are dead. He tried his hand once at making bootleg whiskey on his Kentucky farm. He washed the leftover runny mash down a sinkhole, thereby destroying the evidence. Or so he thought. One evening at milking time, the cows came in from the pasture wobbling and swaying in a most ludicrous manner. First father-in-law feared that they had some strange disease or maybe had gotten into some locoweed (white snakeroot), but they seemed in good enough spirits. In fact, they acted exactly as if they were drunk. Hmmmmmm. Investigation proved that the whiskey slop had finally worked its way down through the hilltop sinkhole and was oozing out of a cleft in the rocks at the bottom of the hill. The cows had bellied up to the bar and sucked the mother lode dry.

Spring Oats

To the cottage husbandman, oats is (if oats seems to require a plural verb to your ear, it is not so spoken among us peasants in rural Ohio) a far more important crop than wheat. It makes good hay and temporary pasture, for one thing. As grain, oats is higher in protein than wheat, and much higher in calcium, iron, thiamin, and fat. It is much lower in carbohydrates. This latter fact makes it an ideal feed to mix with corn, which is high in carbohydrates but relatively low in protein. In fact as I take every occasion to point out, experiments in the 1930s, reported in the previously cited Morrison's *Feeds and Feeding*, indicated that while oats is cheaper than corn to buy, it is worth just as much as corn in a feeding ration where it comprises a third of the ration with corn. Morrison did not seem to know why, and today no one even knows enough to ask why. We're all supposed to grow corn and soybeans and keep our minds shut. Oats is horse feed, and by God, we are up-to-the-minute, forward-looking *entrepreneurialagribusinesspeople* who don't intend to stare into a horse's rear end all day, again. Ever.

So if you are buying feed, a ration of corn and oats is cheaper than corn alone (or corn and wheat—wheat costs more than twice what oats costs) but is possibly just as effective although don't expect the corn and soybean industrialists to ever admit that. If you are raising oats to feed with corn, you can figure your oat crop is worth just as much as your corn although it sells for about a dollar a bushel less at market prices.

That reminds me of something else the cottager should know about. The price you get for selling your grain at the elevator is lower than the price you will pay for buying grain at the elevator. They call it handling charges. The elevator will pay you, at the moment, about $1.40 a bushel for oats. If you come back the next day and want to buy oats, it will cost about $1.80 a bushel if you're lucky. That is one sumbitch of a handling charge. No, life is not fair. So when someone tells you that it is more economical for a cottager with only a small number of animals to buy his grain rather than raise it, remember to figure in this additional cost over market prices when you do your own calculations. Or buy your grain direct from another farmer who won't sock you with a handling charge.

"Mud in the oats and dust in the wheat," as the traditional saying puts it, is more often the truth. Oats is planted as early in spring as the ground will allow, where corn grew the previous year. Disking once or twice is all the pre-planting cultivation necessary. Don't get impatient and try to disk before the ground is sufficiently dry. This is where good drainage becomes particularly advantageous, because it is essential to get oats for grain in the ground as early as possible. M.G. Kains in his classic *Five Acres and Independence* (Greenberg, 1940) claims that oats will sprout on ice, but I rather believe he meant that the grain can be sown on an icy field and sprout after the ice melts and the soil warms up. I must be sure the disk is set to do a very level job because with all the cornstalks on the field, I can't pull a harrow behind the disk to level the soil. The harrow would plug up with stalk residue. I plant oats just like I do wheat and at the same rate, two bushels to the acre. After I plant both oats and accompanying red clover, I may run the disk over the field again, lightly, to level the seedbed and cover the seed. I might finish off the planting by firming the seedbed with the cultipacker.

Because I grow about twice as much oats as wheat, only half of the oat field will rotate to wheat in September. So on that half, I do not plant clover since it would not have time to grow much before I disked the field for wheat.

On the other half, the oats and clover make a sort of wary truce when growing together. In a way they help each other. The legume puts nitrogen in the soil and the natural herbicide emitted by the oats does not faze the clover (it mostly inhibits growth of some grasses). But like seedling trees in the forest, the oats and legume also compete with each other for sunlight. In fact a heavy oat crop can almost choke out clover and especially alfalfa. So, the part of the field that I interplant to clover is the part that I cut early for oat hay as described below. Relieved of the shade cast by the oats earlier than would normally be the case if I waited until the oats matured for grain, the clover rebounds with extra vitality. If weeds have come into the field, mowing the oats for hay also cuts the weeds, preventing most of them from going to seed. The weeds at this stage make fairly good hay too. Making oat hay proceeds just like making legume and grass hay (see the next section).

Meanwhile the other half of the oat field grows on to maturity. If you are like me and can't stand the sight of a big old sourdock growing above the oats and putting on a head with a zillion seeds in it, you have only half the original field to patrol and behead the intruders.

Oats ripens a week after wheat harvest. After combining the grain, I rake the straw into windrows, as I do the wheat straw, scoop up the windrows with the front end loader on the tractor, and store the straw in the barn or a stack next to the barn, for bedding.

Oats of course makes good human food, too. The grain must be hulled, and kitchen-sized hullers for oats do not exist as far as I know. However, you can replace the stationary grinding plates in some hand-operated grain mills with a ⅛-inch thick piece of gum rubber to turn the mill into a huller for rice, spelt, oats, and barley. When moving, the abrasive disk rubs the grain against the rubber and the hulls come loose. Screening the grain is then necessary to remove hull parts. You can get more information from the Corona Company, which makes kitchen-sized mills, and I presume from all small mill manufacturers by this time.

Another way to dehull these grains in small amounts is by parboiling, as is done in the Orient with rice. I haven't tried it, but if it works for rice, it should work for the others. Soak the grain overnight or until it is soft, then cook the wet grain in a pot with about an inch of water in it until steam is rolling off the grain. Dry and then grind in a handmill and sift. The hulls separate much quicker and easier. Then cook the groats.

Specialized commercial millers have oat hullers, practical for commercial growers selling to specialty food markets. Ken West, the enterprising Montana organic farmer mentioned earlier, raises organic oats for oatmeal and ships his grain to a mill in another state for dehulling. The groats are then rolled or flattened for oatmeal.

Some oat lovers try to find a way out of the hulling dilemma by growing hulless or naked oats, but the variety I tried ripened very slowly and unevenly and was a poor yielder. The birds flocked to it in droves and ate or knocked over half the crop.

Making Hay

Haymaking is the most important job on a sustainable farm that raises animals. The first reason is that the crop rotation won't be fully economical without growing hay crops and returning to the land the fed hay as manure. Secondly, good hay is so expensive to buy that it will be difficult to show a profit on sheep or cattle if you do not make your own.

If you buy poor hay in an effort to economize, you are throwing money away because such hay has no value other than as fiber. Sawdust would be better. Poor hay is hay that rain has fallen on after it is cut.

The stems are strawy brown instead of green. It has few leaves left on it, and they are brown too. Good red clover hay is not as green as good alfalfa but more of a dull olive color. Good hay has a softness to it; the stems are not rind-hard. When I have badly rained-on hay, I don't bother to bring it to the barn, but shred it with the rotary chopper-mower and leave it on the field as mulch.

Another kind of poor hay is one cut too late in the plant's growth cycle and dried too long in the windrow. It is hard-stemmed with few leaves and little protein value. A legume gives you the best balance of quantity and quality if cut right after it starts to bloom. But if you must wait because of threat of rain, the hay will still be nutritionally superior to rained-on hay even if cut a few days before or as much as ten days after it reaches full bloom. If you mow hay too far in advance of blooming, it will be so green that it will take a day longer to dry—a day more with risk of rain. One advantage of making hay after it is a little too mature is that it will dry quicker.

Clover or alfalfa hay, alone or intermixed with timothy or some other tall grass, is the mainstay of the husbandman in winter. If legume hay is put up correctly it can be the only feed you need for ruminant animals. Even pigs and chickens can derive a third of their food from forages. If you can't raise both your own grain and hay, it is, at current prices, cheaper to raise the hay and buy the grain.

Haymaking involves the same procedure whether it is grass hay, cereal grain hay, or legume hay. Essentially, the plants are mowed with a sickle bar mower (I have also used a rotary mower on meadow grass hay with good results even though the rotary mower shreds the hay considerably), allowed to dry in the field, raked into windrows, and then baled and barned or hauled "loose" (unbaled) to the barn or stack. The trick is to cut hay when there will be at least three days of sunny weather in a row. I study the weather map on television with an eagle eye, hoping to spot a three or four day spate of dry weather. I know from experience that the meteorologists are very good at giving today's weather, and fairly good at giving tomorrow's weather. Beyond that, I can do about as well with one eye on the sky and one on the weather map.

The disadvantage of oat hay is that it takes at least a day longer to dry than clover. The advantage is that we are talking July 1, usually, for oat hay, and the chance of getting a week of dry weather then is better than in June, the major haymaking month.

Grasses, of which oats is one, should be cut right as they form a seedhead or just before. If you cut them after the seedheads mature, the plants will have stored their protein in the grain and the rest of the hay will have little taste or nutrition. I try to cut at milk stage when the grain is partially developed but the stems are still green, hopefully getting the best compromise between nutritious green matter and nutritious seedheads.

I leave the hay in the swath until it is about two-thirds dry, or for about two days of sunny weather, then with a side delivery rake and tractor, windrow the hay and let it dry until it is safe to pile in the barn loft without heating up and starting a fire. The raking requires artfulness as well, not so much with grass hay like oats, but with legumes. If the clover and especially alfalfa gets too dry, the rake knocks too many leaves off. If the legume is raked too wet, drying in the windrow will proceed too slowly. In dry weather, I like to rake in the morning when some of the night's moisture is still in the hay to decrease leaf shattering.

I can't describe with words how you will know when the hay is dry enough to put into the barn. Only experience can guide you. I can tell by the sound of the rustle the hay makes when I fork it on the truck and by the heft of a fork full. I can tell by how easily the fork slips into the hay, and how easily the hay comes off the fork. When the hay is still too wet, handling it is like trying to lift forkfuls of wet rags.

Putting a large amount of uncured hay in the mow is a good way to burn a barn down, but since we make only small amounts at a time, I can spread *slightly* moist hay out over the whole loft, about two feet deep, and eventually the hay dries without heating up too badly. Sometimes a little mold develops in it, but it seems to feed all right. I worry sometimes about breathing the moldy dust that puffs into the air when I fork that kind of hay down from the loft to the animals, but we all seem to have survived so far.

There is so much more technology to haymaking that has not been discovered or rediscovered, especially for small cottage operations. The present practice of compacting hay into some kind of bale, especially the large round bales which are too cumbersome for cottage farmers and too wasteful of hay, may not be the best way for the smallest farms. New-man Turner in his book *Fertility Pastures and Cover Crops* (Faber and Faber, Ltd., 1955; reprinted by Bargyla and Gylver Rateaver, 1975) de-scribes more sophisticated stacking methods that allow a farmer in very humid regions to put hay soon after it is mown in small stacks (or doo-dles, as my grandfather called them). These doodles are supported by lit-tle tripods of poles instead of a single pole of common tradition. Once in the stack, the hay is not much harmed by rain as it would be lying in a windrow and eventually the stack dries out without molding because the tripod affords good air circulation.

With Turner's method, mown hay is picked up out of the swath or windrow and delivered to these little stacks in the same way we moved it to the large stacks of my grandfather's fields: with what is called a buck-rake, a large wooden forklift, in Turner's case attached to a tractor, which scoops up a fairly large amount of hay at one time. After the hay dries, the buckrake can move the little stacks to the livestock as needed.

Reading Turner's book a year ago, I realized that whether or not I ever used his stacking method, I could surely copy the way he moved hay from the windrow with my tractor's front-end loader. My son, the wood wizard, built his version of the buckrake in one day out of our own white oak. Instead of laboriously hand-forking hay into the pickup truck and then hand-forking it even more laboriously from the truck into the second story of the barn, where Carol would hand-fork it on back farther into the loft, I now use the tractor and "buckrake." Lower-ing the big wooden fork into the windrow of hay, I simply drive along until the fork is full, raise it a little, drive to the barn, raise the forkload to the loft door, and then standing inside, hand-fork the hay off and stow it away. Three hand operations reduced to one. I can make hay faster than previously, and do it alone.

Now I realize that I can stack hay outdoors with this rig without *any* hand-forking. This practice, if perfected for use on cottage farms univer-sally, would mean saving money in a big way by limiting the need for indoor hay storage. I think loose hay outdoors cures better than baled hay anyway, and today plastic film is available to cover the stack top so it sheds water well. With wood bunkers around the stacks, animals could self-feed from them except in spring-thaw time. Also, being outside most of the time, the animals would spread most of their winter manure themselves. To alleviate the mud problem of congregating around the stacks, I would locate the stacks on the highest, best drained spots in the field, where the soil needs the heaviest deposits of manure anyway. I can hardly wait to try this scheme out.

Many nineteenth-century farms operated this way, sometimes even building frames over which the hay was stacked, so that the animals had some shelter during the cruellest weather. For very small farms that can't justify the high cost of machinery or labor these days (can any farm justify it?) these old methods are again economical and practical. Perhaps they always have been, and we have only been fooling ourselves with the "advanced" technology of the twentieth century.

To hurry along hay drying, most farmers use what is called a hay crusher or conditioner. The conditioner, pulled by the tractor, sometimes directly behind the mower, pulls the hay from the swath through a set of rollers which crush or break the hay stems with an action like that of an old wringer washing machine. The breaking of the stems allows them to dry out faster. The conditioner is another piece of equipment I have been able to get along without on my small farm, but I found it advantageous for our hundred cow dairy farm in earlier years.

In the kind of rotation I follow, where there are five fields, one each of corn, oats, wheat, first full-year hay, and second full-year hay or pasture, I will possibly have hay to make in all plots except the corn. The wheat plot's seedling clover can be made in September after wheat harvest, as can the seedling growth in the oats field. Two cuttings each can be made in both the first year and the second clover plots (three cuttings with alfalfa). Also there may be an early cutting of grass hay from the pastures during the flush of June growth. The second-year clover plot's second cutting can be left for pasture into the late fall. The first year's second cutting can be harvested for clover seed instead of hay.

Southern graziers have developed to a very high degree the art of using oats for hay or rotational grazing, especially for winter pasture. Northern graziers are only just learning because it has always been believed, until the techniques of intensive grazing came along, that winter pasture in the north was not practical. We now know that red clover and some of the grasses, like fescue, can be grazed under controlled condi-

213

tions whenever there's no heavy snow on the ground and mud is not a problem. Even after the hay and grass have died and turned somewhat brown (like hay, in other words) the plants have as much or more nutritional value as hay.

Oats can be sown for fall and winter pasture or hay after an early sweet corn plot has been harvested and the stalks grazed down by livestock, or cut and hauled to the livestock. Rye grass can also be used this way, but the husbandman who is raising oats anyway is dollars ahead just using his own homegrown product for temporary pastures.

Beans and Minor Cereal Grains

You can play all kinds of tunes with the marvelous synergy of crop rotations. Let us suppose that drouth kills the seedling-clover stand in the wheat. After wheat harvest, I can resow the clover. Or I can plant oats for late summer pasture. Or I could plant buckwheat for grain, or soybeans and other beans for hay or for human food.

Soybeans are a staple for vegetarian farmers. Most years they can be sown as late as July in the north and still produce a crop of beans before frost. This is the great advantage of beans in crop rotations. Two years ago, when it became apparent in June that drouth had ruined my new seedling clover, I made what hay there was, plowed the field, and sowed soybeans. Not wishing to use herbicides I planted the beans in rows like corn so I could cultivate them. I used my double garden seeder for planting, just as for corn, and used the string bean plate, which planted the soybeans perhaps a little too thick. But they grew beautifully, well above my waist height.

Soybeans being a legume can draw nitrogen from the air for fertility and so, like all legumes, are an excellent crop for organic farmers. For the beekeeper like me, soybeans are also a source of superb honey.

Navy beans and similar dried beans can be planted the same way. All beans can be harvested with a combine, just as you would cereal grains.

You should have seen me haul in my 26 bushels of soybeans to the elevator last fall. At the height of the harvest, ton upon ton of beans were being brought to market, and I found myself in my pickup wedged between two semi-truck loads in the long line of trucks and dump wagons waiting to unload at the elevator. I was treated something like the village idiot—everyone was glad there was still a smaller farmer around to feel superior to. When the test station operator lowered the big suction pipe into my pickup to get a sample of my beans, the other farmers teased me good-naturedly, pretending to worry that the pipe would suck the whole 26 bushels and the pickup as well—whoosh!—into the test bucket before the operator could turn the suction motor off. Yuk yuk yuk. But I got $5.85 a bushel for my half-acre of beans, the same price everyone else was getting, and that $152.10 represented about five hours of work and a little gasoline, or about $28 an hour return on my labor—a whole lot more than the mega-farmers' per-acre profit. I also took off some hay from that half acre before I planted the beans, which normally (if not played out because of the preceding year's drouth) would have been a ton and a half of clover hay worth $150. And after I combined the beans, I planted wheat and grazed the wheat in November.

I keep thinking of intriguing new tunes to play on the rotation melodeon. What would happen if I planted an heirloom table bean like Jacob's Cattle (which would sell for a great deal more than soybeans) after a cutting of hay, and followed by wheat for grazing and grain? I now have ten trillion and one projects on the agenda.

Rice is the staple cereal grain in the Orient, where, it should be remembered, the cottage farm is the premier kind of farm. As Richard Critchfield points out in his *Trees, Why Do You Wait?* (Island Press, 1991), in my opinion the most accurate account of the decline of rural society in America: "Despite huge crop increases, farms in India and China are not getting bigger and fewer the way they are in America. Indeed, the number of draft animals in both countries has grown." Most of the world's populations in the Third World countries are in fact fed by cottage farming. Historically, small Chinese farms have fed more people per acre than the mightiest, modern American mega-farm does. Even in very industrial, first-world Japan, the average-sized farm is barely more than four acres, and is it not true that Japan's economic health and growth is better than ours? An American anthropologist, who happened also to be a gardener, told me once how amazed he was to find cottage farms in New Guinea that both in variety and amounts of food out-yielded industrial agriculture three to one.

Barley is an important grain for areas like Montana where corn does not grow well, since livestock and hogs can be fattened on it. Northwestern cash grain farmers not focused on livestock grow it in place of oats because they believe they can make about $20 more per acre with it. Also what would beer drinkers do without malt barley? Barley is grown exactly the way wheat is grown. Sometimes it is hard to tell the two apart in the field from a distance.

Rye adapts to the most northerly grain growing regions. It will germinate at colder temperatures than any other cereal grain. Someone needs to grow rye for rye bread and good Scotch whiskey. In the latter case, leave it to the Scots, I say, who have the climate and know-how for rye. If you want rye flour of your own, then plant some rye. Otherwise, for cover cropping and temporary pastures, oats is better, especially if you are planting oats anyway: you don't have to buy more seed.

Spelt is related to wheat but has one notable difference. It is very low in gluten. For people who can't eat conventional bread because their systems are intolerant to wheat gluten, spelt is about their only choice. Stan Evans Bakery in Grandview, Ohio, near the heart of the spelt producing regions of eastern Ohio, sells a lot of spelt bread. Farmers who traditionally grow spelt say it is better livestock feed than other cereal grains. Although that claim is often disputed, recent research seems to indicate that spelt aids the digestion of other foods. Grow it just as you would grow wheat.

Triticale is a modern cross between wheat and rye and has been hailed as a miracle crop to solve all our problems except how to propel the Cleveland Indians into a World Series. Southern plains ranchers, who sometimes see their wheat wiped out by mosaic diseases, claim that triticale makes a flour as good as wheat flour, and a forage crop (for hay or grazing) better than wheat. Triticale is grown just like wheat. For my climate, wheat is better. A neighbor grew triticale. Once.

Buckwheat is another "grain" that some organic farmers grow for a cash crop. The market is usually in oversupply. Buckwheat's uses as a cover crop and weed choker are usually exaggerated. If you like buckwheat cakes, grow a little as a garden crop.

The best source of seed for all kinds of grain is The Grain Exchange at The Land Institute (2440 East Water Well Road, Salina, Kansas, 67401). The Grain Exchange also publishes good information in its newsletter on ways to harvest and clean grains, and process them into flour.

The Land Institute is an interesting story. Its founder, Wes Jackson, another of my favorite farmers—and certainly the contrariest of them all—became bored and impatient with conventional university approaches to agricultural research, left a career as a plant geneticist in the warm, secure womb of the University of California, went home to Kansas, and started his own school and research farm, the Land Institute. On his own, Wes has become probably the most well-known and certainly most written-about plant geneticist of today. The only magazines that steadfastly refuse to interview him and profile his work are the commercial farm magazines, which should have been the first to recognize and encourage his genius.

Wes has more vision than a whole think-tank full of crystal ballers. He talks with the crackling dry wit of the Plains and says naughty things. Considering the excesses of technology, he grins wickedly and remarks: "Hell is now technologically feasible." Asked his opinion of modern education, he flashes his toothy, mischievous smile and observes: "For tens of thousands of students, the universities have become little more than holding pens that keep them off the job market where millions of hours are devoted to turning out work too shoddy to be either useful or artistic. What we need is not more college majors in upward mobility but a major in homecoming." See his latest book, *Becoming Native to This Place* (University Press of Kentucky, 1993). Jackson's vision (one of many visions) is a wholly sustainable complex of cottage farms and villages based on a permanent prairie agriculture, each community drawing its energy from the sun, producing its food and fiber needs with little outside reliance on fossil fuel technology—a biologically sophisticated prairie version of my meadow farming. To that end, he and his staff and students have been growing combinations of perennial prairie plants with exotic names like Illinois bundleflower,

Maximilian sunflowers, and sideoats grama, which have the potential today to produce food as well as annual grain crops without cultivation, without heavy machinery, and without chemical fertilizer or pesticides. Walking on his native prairie beside his research plots, Wes is overcome with excitement. "There are two hundred species of plants in a typical square mile of this native prairie," he exclaims. "All that information growing there, begging us to draw it out. Add to that the vertebrate animals, the algae, fungi, insects, and all the vast numbers and diversity of micro-organisms that live among these plants. Add to that the diversity within the species. The amount of biological information is just awesome and we know only a small amount of it."

Harvesting Grain

The practical difference between corn and the cereal grains is the size of the ear. Corn or maize ears are large enough to make hand-harvesting of a small acreage practical. Small grains are practical to harvest by hand only in small garden plantings. We used to thresh out about a bushel of garden-grown wheat for our own bread (a bushel makes about sixty loaves) using toy plastic baseball bats and then winnowing the grain in front of a large window fan.

For larger amounts some kind of mechanical way to thresh is necessary, and therein lies the problem for cottage farmers. At this stage in history, nobody is making a reasonably priced cereal grain combine for the small farmer. Vogel seed plot harvesters used by university research stations are expensive (paid for by the taxpayer, of course). So are the small harvesters made in the orient for the preponderantly cottage farm culture of China and Japan. Large farm combines are even more expensive. A new one with a sixteen-foot header now costs nearly $200,000, which farmers couldn't afford if the government did not heavily subsidize cash grain farming. (In a way, it is unfair to criticize farmers for taking these subsidies, because mostly the money goes right through their hands to pay the salaries of the people who make this expensive equipment.) Old, smaller combines like mine are still available cheap, but not easy to find in working order (see chapter 9).

The best news for small farmers on the grain harvesting front is the invention, or re-invention, of the *stripper harvester* of ancient origin.

This new version uses a rotary brush and fixed teeth to literally strip seed or grain out of the heads and then suck them by vacuum into the hopper. The stripper fits conveniently on any tractor front-end loader. A 50-horsepower tractor is big enough. The strippers sell as low as just under $10,000. That is still hefty for a cottage farmer, but nothing else today in new field combines even comes close to being that low-priced. The strippers are made primarily for harvesting fine grass seed, but the manufacturer says they will harvest cereal grains too. Some chaff might remain with the grain but this would not necessarily be a problem for the cottage farmer intending to feed his grain on his farm. The company is Ag-Renewal, Inc. (1710 Airport Road, Weatherford, Oklahoma 73096). If a cottage farm market developed for these harvesters I'm sure they would get cheaper and most likely be improved for cereal grain harvesting.

Of course, threshing and winnowing only 50 to 100 bushels of grain need not be a horrendous job using old traditional methods. Before threshing machines (and still today in some Third World countries like Africa) bunches of grain stalks were laid out, usually in a circle, on a clean barn floor (or hard clay or pavement) and a horse was driven round and round, the trampling action of its hooves knocking the grains out of the heads. The straw was then stacked for use as bedding and the grains swept into piles. On a windy day, front and back doors on the barn floor were opened, and the grain winnowed in the strong air draft that funneled through the barn. When human labor was plentiful, the threshing was more often done with flails, which in a skilled hand did a much better job than horse hooves.

There's said to be an easier way to thresh than using horses' hooves, though I have not tried it myself. If you had a large, smooth, clean floor (barn or otherwise), instead of running over the grain stalks with a horse, use a lawn tractor. The rubber tires will knock the ripe grain out without injuring it. The grain can then be swept up and winnowed any way you can capture a good breeze. If you think the grain is not clean enough after this weird threshing method, you can wash it, a little at a time, when you bring in a batch to grind and bake into a loaf of bread, running water quickly through it in a colander then spreading the grain out in the sun on a clean sheet so it dries quickly.

I am not really advocating the hand-threshing of grain, except for a few bushels for the human family, but better that we know it can be done than to starve to death some day for lack of a combine. Obviously, in the interim, a combine is the answer and if you can't find one you can afford, you might be able to hire a neighborhood farmer to harvest your grain for you. If you have a larger cottage farm with, say, ten to forty acres of grain, you will find that in most cases it is cheaper at present to hire out the combining than to own a combine—unless you are an excellent mechanic and enjoy restoring old machinery.

There are other alternatives. In Amish country there are plenty of old binders and stationary threshing machines around. Really tenacious cottage farmers could equip themselves with both. A binder cuts the grain and binds it into bundles. The bundles have to be set up into shocks to complete the ripening and drying process. Then the bundles are run through the thresher, the straw blown into strawstacks and the grain separated and augered into a grain wagon or truck.

It was my good fortune to have been living in a "backward" area of Minnesota where communal threshing was still in style in the 1950s. We would go from farm to farm, following the thresher, and harvest each farm's grain. The thresher that served the five farms in our "threshing ring" cost $2000 new in the 1930s and so the annual cost, including repairs to keep it running, was hardly $100 a year. Compare that to today's $200,000 combine which, even while harvesting two to three times more acres a year than those five farms comprised, will cost at least a hundred times more and wear out sooner. That's just another example of why the Amish are often rich while their mega-farm neighbors are dependent on subsidies to make a profit.

But what I cherish now, looking back, is not the economy of communal threshing, but the fun of it. Fifteen of us, more or less, working and joking together in the fields and eating five (yes, five) scrumptious meals rich in cholesterol every day without a qualm. Families bonded together, whether they liked each other or not, in a common economic interest: getting the crop in. That was a real community. What we share today in the pale light of the television screen is something else.

I discovered last year another alternative to the combine—on the Kuerner farm mentioned earlier, where the most significant crop is the masterful paintings of Andrew Wyeth. Karl Kuerner, Jr., found a few years ago a cagey way to harvest oats. He has a hay baler because he makes quite a bit of hay, but for the few acres of oats to feed to horses and cows, he couldn't justify a combine.

Karl reasoned thusly: Only cows and horses are going to eat this oats. Cows and horses don't care if the oats is hulled out or not. In fact cows and horses love to eat oat hay and even oat straw as a fiber supplement to grain. (The Minnesota farmer I worked for years ago said in his year of greatest drouth, he kept the cows alive on oats straw, all the feed that was left.) Karl decided to let the oat grains ripen just a little more than when making oat hay, then simply cut the crop as if it were hay, windrow it, and bale it. In the winter, the bales could be broken up in the mangers and the livestock could munch the oat grains out of the straw or along with some of the straw. What they didn't eat could be thrown out of the mangers and used for bedding.

Of course it worked. Why wouldn't it? If balers had been invented before threshers, maybe that's the way livestock farmers would harvest all their oats for livestock today.

When my wife and I had a two-acre garden but no farm, we used a variation of this method with wheat. We cut the wheat with a scythe, raked the stalks together in bunches, and crammed the bunches into the little loft above the chicken coop. Every day I'd pull down a handful or two of stalks and the chickens would greedily eat the grains out of the heads. The straw became bedding.

Today I do the "Kuerner maneuver" with my own oats, only not with a baler. I make hay out of some of the oats, and since oat stems dry slower than clover stems, I do not mind if some of the oats is approaching maturity when I mow it because drying will be faster. Then in winter, I throw the oat hay down just like the rest of the hay, and how the sheep go after those dried-in-the-milk oat grains. They seem to relish the oats more this way than when I feed them mature grain in their grain trough. Maybe I don't need my combine at all.

The only drawback I see to this method of harvesting grain is that if the grains are mature in the straw, they might draw a plague of mice and rats into the barn. So far, with our cats standing by, this hasn't happened.

Storing Grain

Storing cereal grains requires considerable attention to details. The word, *granary*, has almost gone out of use, but when I was a boy, it was as common as the word "car" today. The granary was a rather small building that commanded the center of our barnyard. It contained five grain bins, four of them about ten-foot square and the back room about ten by twenty, all able to be filled to a height of about ten feet. An aisle from the only doorway in front gave access to all five bins, two on each side and the bigger bin at the end of the aisle. The bin walls were all wood. Unlike metal, wood "breathes" and is more absorbent of moisture. If grain were not quite as dry as it should be (13 percent moisture or less is safe), there would be less risk of mold in a wood bin than in a metal one.

The floor of the building was raised so that it would be on about the same level as a truck or wagon bed backed up to the doorway. Sacks of grain or ground feed could then be moved on and off the beds more easily. But there was another practical reason that the granary floor sat well off the ground. Dogs, cats, chickens, even humans if necessary could get into the crawl space under the building. This greatly discouraged rats and mice from making their homes there. The supports upon which the building sat were originally sheathed in tin to further discourage rodents from climbing up and gnawing holes through the floor.

Shuttered window openings were built into the wall above each bin, so that grain could be shoveled in from wagons pulled alongside the granary. The inside aisle "doors" of the bins were made up of individual boards that fitted into slots on each side of the opening, and were added or removed as the level of grain in the bin rose at harvest and then fell as the grain was fed. Before new grain was added to any bin it was cleaned out and treated with insecticide to kill all weevils that might be present (malathion was, and is, effective, and safer than fumigants used today). A standing rule was never to allow old grain to stay in a bin over one year. Weevils would be more apt to become established in that old grain and would be exceedingly difficult to exterminate even with fumigation. Although highly toxic fumigants are used widely in commercial grain storage apparently without harming consumers, I am glad I eat mostly my own grain. What we eat, about a bushel a year, we store in the freezer which insures no weevil infestation. We thoroughly clean out and spray with household Raid the bins where the animal's supply is kept. If you do not want to be an organic heretic like me, you can try diatomaceous earth, the prescribed organic remedy. Be sure to get the insecticidal kind, not the kind used in swimming pool filters. If diatomaceous earth doesn't work for you, don't blame me.

One of the sins of my young and foolish manhood was to tear the granary down. The building was still in perfectly good shape, but it did not fit into our new and glorious plans to become large-scale, agribusiness dairy farmers with a hundred cows. The granary was not large enough to handle that much grain and so it had to go. My granary today is much smaller than the one I tore down. It consists simply of two bins made of plywood, built into the front of the corncrib. I also store oats and wheat in steel barrels if I know the grain is very dry.

I describe the fine points of a traditional granary because, though "obsolete" in modern farming, it is still most suitable for the cottage farm, the size and number of its bins depending on the number of animals on the farm. In it can be stored any of the cereal grains: oats, wheat, rye, barley, buckwheat, and I suppose rice although I have no experience with the latter. The farmer could store in his granary any small-sized seed grown and harvested in the same general fashion as the cereal grains, including soybeans, dried beans like navies, grain sorghum, sunflowers, amaranth—just about everything except corn, which on the cottage farm is, or should be, harvested on the ear and stored in airy cribs.

I have not said much about straw but this, too, is an essential "crop" for husbandmen. You must bed your animals with something, and while I understand that shredded paper is okay for bedding now that we need to recycle everything, paper isn't free and your own straw more or less is. Instead of having to buy bedding, you have your own as by-product to the grain, and then you haul it back out to the field, laden with manure, for the best fertilizer in the world.

Try to haul your manure in late summer or fall, not on frozen ground, so that when rain falls, the nutrients soak readily into the dry soil. Or spread it on the old hay field right prior to plowing it under. If you have a good spreader that breaks the manure up and is able to spread it thinly, applying it to the new seedling clover after wheat or oat harvest not only gives nutrients to the legume but acts as a moisture conserving mulch. It can be applied to newly seeded corn fields or wheat fields as well for the same reasons, and to alleviate crusting on clay soil, should a hard rain fall before the seeds come up.

Having finished that job, you have completed the annual cycle of the grain fields and can sit on a hill in the last mild days of November and watch the wheat field grow green until it sleeps beneath the snow. If I have a wish, it is to echo the words of the heroine in Mildred Walker's wonderful novel of 1948, *Winter Wheat*, as she thinks about how full of suicidal despair she had been when news of her fiancé's death in war came to her, and how finally the farm and the promise of the wheat fields of her home lifted her spirits: "How had I ever felt that way? I didn't feel that way any more. Now I wanted to live my life with the strength of the winter wheat, through drouth and rain and snow and sun."

Books the Contrary
Farmer Treasures

The most difficult problem that an agricultural writer faces is convincing readers that farming is very much a part of the whole societal structure, not just another job by which a person makes a living. Therefore, what happens out in the countryside in one century invariably affects the cities in the next century, and often there is not even that much of a time gap. Food is the common denominator of life; producing food is part of the biological and cultural as well as economic fabric of a civilization. Although historians should continue to question the precise causal connections between rural and urban economies, it is a fact that a strong and vital urban society has always been supported by a strong and vital rural society. General decline in the Roman empire, the British empire, and the Russian Communist empire paralleled if not followed the decline of their rural communities. The same thing is happening in the United States, in my opinion, only we don't realize it yet. The following books all reflect my hypothesis in one way or another.

Rural Rides, by William Cobbett (last published by J.M. Dent and Sons, 1973, in the Everyman's Library Series) is still in print. Any book by William Cobbett will do, but in my opinion this one is the best for pointing out what happens when political greed ignores the plight of rural people and allows industrialism to overwhelm pastoralism. Cobbett is my role model although he was much more courageous than I could be, going to jail rather than quit his acid and fiery criticism of England's rape of its farm workers and the country's slow descent into centralized totalitarianism. Cobbett, active in the first half of the nineteenth century, accurately predicted England's decline as a world power and accurately pointed out, incessantly, its cause: impoverishing rural areas and forcing farmers and craftspeople to adopt the machine philosophy of the industrial revolution. Cobbett was also the first writer I know to question the practicality for formal schooling when educational information was readily available from other sources. He recognized that

225

formal schools were little more than the tool with which people in power persuaded children to accept that power. Along with his angry common sense, Cobbett was an excellent writer, so reading him is fun. However, he is writing in the immediacy of the conflict in which he was embroiled, and to understand him, the reader must have a grasp of English history from the eighteenth to the twentieth century. Go ahead. Make that a project. You will find yourself reading a history of a country's descent from a world power to a third-rate power that is eerily and frighteningly like the description of what is happening in the United States.

I recommend all of Wendell Berry's books. I consider his *Farming: A Hand Book* (a Harvest/HBJ book, Harcourt Brace Jovanovich, 1967) to be more important, more telling—literally as much a handbook of agricultural reform as it is a book of wondrous poetry—than his essays, though his essays are magnificent. To a greater degree than any other writer I have read, Berry shows how the attempt to separate farming and the general society culturally or practically does great harm to both. His books are wonderful moral treatises, really, in an age where journalists are taught to avoid the moral aspects of what they are trying to write about—as if that were possible.

Berry's most recent books are published by Pantheon. Earlier books published by North Point Press are now available from Farrar, Straus, and Giroux. Probably his best-known book, *The Unsettling of America*, was published by Sierra Club Books. My second favorite, after *Farming: A Hand Book*, is *The Long-Legged House*, publishing by Harcourt, Brace, and World in 1969. The chapter "A Native Hill" moved me to return to my homeland and to farming.

Wes Jackson, farmer, plant geneticist, and now trying to rebuild a dying village in rural Kansas, is to my notion the most original thinker in American agriculture today. He seems to get progressively more penetrating and creative in his analysis of how to achieve a sustainable civilization as he goes along. And Jackson does not just talk. He acts. His latest book is *Becoming Native to This Place* (University Press of Kentucky, 1993).

Akenfield, by Ronald Blythe (published in 1969, available from Pantheon, as part of its Village Series), is a profile of an English village. The book makes a thoughtful Afterword to Cobbett a century later, and helps one to understand the difference between a rural village and a larger town or city. It is also just plain interesting reading about the rural mind. The book is a series of interviews with various residents of Aken-field in the south of England, and is very, very genuine, although Blythe cannot quite rid himself of the usual urban prejudice that a person who chooses to stay in a village or on a farm throughout a lifetime will suffer from cultural narrowmindedness—as if moving around in cities will cure that fault. Many of his interviewees repudiate this prejudice soundly, but Blythe never seems to be quite convinced.

A raft of books has been published over the last seven years or so, attempting, in the conventional journalistic manner, to describe the present state of commercial agriculture and the decline of traditional rural society. These books are mostly good, but I think that Richard Critchfield's *Trees, Why Do You Wait?* is the most telling (Island Press, 1991). It focuses on two communities, one in North Dakota and one in Iowa, but the situations he describes are very similar to those here in my part of Ohio.

But rather than broad journalistic reportage, however well researched, I think that personal glimpses and private meditations of the solitary pursuits of country people within their own lives reveal more about real farming and real country life. I am fond of John Baskin's *In Praise of Practical Fertilizer* (W.W. Norton, 1984) and his earlier *New Burlington* (also from Norton). I draw great satisfaction from David Grayson's old books, especially *Adventures in Contentment* (originally published in 1907, republished by Renaissance House, 1986), remembering that Grayson, whose real name was Ray Stannard Baker, wrote his country books as an aside to a career of courageous and controversial journalism. Harlan Hubbard's *Payne Hollow* is another book I love very much. I was fortunate enough to visit Harlan at his homestead on the Ohio River twice before he died. He was the most authentic of contrary farmers. Harlan and his wife Anna lived quite elegantly without electricity, telephone, or automobile.

I assume that following all the journalistic endeavors, there will be another wave of novels about the decline of traditional rural life. I am writing one myself, as a matter of fact. What I have seen currently in this genre is way too depressing for my taste or philosophy. The subject is one of sadness, to be true, but I'm convinced that most farm novelists left the farms they feel compelled to write about because they despised farming, and the sadness they exhibit now is hypocrisy: they turn sorrow into a tool to slyly belittle farm life, thereby in a backhanded sort of way justifying its demise and their flight from it. I prefer tough but loving portrayals, spiced with wit and humor, like *Smith and Other Events* by Paul St. Pierre, which is set in western Canadian cattle and homestead country, and rings with genuinity (Beaufort Books, 1984). None of that pathetic oh-dear-me attitude as exhibited in Jane Smiley's award-winning *A Thousand Acres*. A chapter in Louis Bromfield's *Pleasant Valley*, titled "My Ninety Acres," is my favorite short story. This portrayal of the love of an old man for his farm is so authentic that for years I thought it was a true account, not fiction. *The Thresher* by Herbert Krause (Bobbs-Merrill, 1944) is another authentically written and well-written novel that comes to grips with great tragedies in farming without becoming maudlin, ghost-worshipping, or nihilistic. Krause loved and respected farmers, you can bet on it. The same goes for Mildred Walker's *Winter Wheat* (Harcourt, Brace & Co., 1944) and Ralph Moody's *Fields of Home* (W.W. Norton, 1953).

I have to mention *Modern Meat* by Orville Schell (Random House, 1984). I often hear defenders of organic farming lament the use of chemical fertilizers, which in moderation do more good than harm, and pesticides, some of which are dangerously toxic and some at least as "safe" as gasoline. But rarely do the organicists talk about medicines, hormones, and antibiotics being used and misused in the livestock industry, where in my opinion, the greatest dangers to health lie. Read *Modern Meat.*

James Agee's *Let Us Now Praise Famous Men*, about poor sharecroppers in the south in the 1930s, is a book I often draw on for inspiration, especially the chapter on education. This short essay, at least in the first few pages, has nothing particularly to do with farming or rural society, but is for me the most penetrating comment on the flawed psychology of modern schooling that I have ever read. I wonder if anyone else, professing to be a teacher today, has read that chapter, or if they have, if they understand what Agee is saying. He is not the easiest writer to read.

Marion Nicholl Rawson's lovely, quaint book of 1939, *Forever the Farm*, is another of my treasures. Rawson describes the life, culture, tools, and architecture of pre-industrial agriculture in America as she could still find it in out-of-the-way places in the early twentieth century. The book contains historically valuable material, but I treasure it most for the rural speech patterns Rawson has preserved and for the little cameo short stories that she sprinkles among the drawings and pages of descriptive and historical information. Probably this is becoming something of a rare book by now. My sister waited until our library put its copy on the discard pile and then she pounced on it and gave it to me as a birthday present.

I seldom find anything of remarkable interest in scholarly books about subjects relating to farming, but David W. Orr's new *Ecological Literacy* (State University of New York Press, 1992) is most helpful in showing the close relationship between sustainable farming, sustainable education, and sustainable societies. Orr's writing style is also clear and sprightly.

Look to the Land by Lord Northbourne (published in 1940 by J.M. Dent & Sons, and quoted at length in this book) is the best articulation of pastoral economics, as distinct from and opposed to industrial economics, that I have read. Unfortunately the book is rare and out of print. It should be required reading in every university course in economics.

For the practical small farmer looking for ways to make money from small-holding, Andrew W. Lee's *Backyard Market Gardening* (Good Earth Publications, 1993) is an excellent source of information on fruits and vegetables.

And finally, everyone should read Jean Giono's little book *The Man Who Planted Trees* (Chelsea Green, 1985), a fictional account of how the restoration of ecological diversity to a plundered, deforested land not only led to a return of wildlife, diverse plant life, and clear running streams, but to the return of healthy human communities and viable, sustainable human economies as well. I don't know of any book that articulates as accurately and as beautifully the fundamental philosophy behind *The Contrary Farmer*.

Index

E

Earthway planters, 43
earthworms, 80–82
economics: industrial, 16–37; pastoral,
  16–37; underground, 32–35
education, costs of, 25–26
eggs, 61–64
egrets, 104–105
Eisner, Thomas, 153
EPA (Environmental Protection
  Agency), 163
equipment, 156, 175–99. *See also* machinery; tools; specific equipment
Erickson, Siri, 102–103
erosion, 112, 155–57
ewes. *See* sheep

F

Fabre, Henri, 4
farm size, 20
Farm Journal, 39, 128
farming: grassland, 105–106, 110–22;
  no-till, 155, 190; organic, 40–43;
  work of, 1–15
Faulkner, Edward, 40
felling trees, 141–46
fencing tools, 192–95
Ferguson, 184
fertilizer, 65, 28, 41–42, 115, 161–63
fescue, Kentucky, 113
Fichtner, John, 123–24
finances. *See* economics
firewood, 127–28, 145–46, 195
fish, 80–81, 90–91
flooding, 84–85, 94–95
flowers, wild, 116, 122–23
food chain, 53–54
Fords, 184
forestry, 132–48

forks, 196
fowl, 61–64. See also animals
French intensive gardening, 42
Freshwater Farms of Ohio, 81
frogs, 98
fruits: *see* produce; specific fruits
fruit trees, 136–37
fuel, wood, 127–28, 145–46, 195

G

gardening: benefits of, 50–52; French
  intensive, 42; and innovations,
  38–52; kitchen, 47–50; and mulch,
  38–40; subscription, 45–46; and
  water conditions, 86–87
Garland, Hamlin, 2–3
geese, Canada, 99
Giono, Jean, 229
Grain Exchange, The, 217
grain harvester, 192
grains, 200–24; storing, 222–23
grain scoop shovels, 196–97
granary, 222–23
grasses, 119–20, 121–22
grassland farming, 105–106, 110–22
Grayson, David, 227
grazing, controlled rotational, 105–106,
  110–22
grease gun, 198
Green Magazine, 184

H

handiness, 2
harrows, 189
harvesting: corn, 151, 169–74; grain,
  218–22
Harwood, Dr. Richard, 51
hay, 105, 115–16, 120, 209–14
hay rake, 191

GENE LOGSDON grew up on a farm in the 1930s. He attended Catholic seminary and several colleges of philosophy and theology that no longer exist, tried his hand at agribusiness dairy farming, and eventually fulfilled the requirements for a Ph.D. in American studies and folklore, but was denied his degree when he refused a teaching position, opting instead for a stint as a suburban journalist writing about country life. Logsdon now lives in Upper Sandusky, Ohio, a mile from his boyhood home, and divides his time between writing and raising fruits, grains, vegetables, and livestock.

Other Books by Gene Logsdon

Wyeth People

Two Acre Eden

Homesteading

Successful Berry Growing

Better Soil

Small-Scale Grain Raising

Getting Food from Water

Wildlife in Your Garden

Moneysaving Secrets

Organic Orcharding

Practical Skills

The Low Maintenance Home

At Nature's Pace

Related books from Chelsea Green

(In addition to the Real Goods Independent Living Books already listed.)

The Man Who Planted Trees by Jean Giono

The New Organic Grower: A Master's Manual of Tools and Techniques for the Home and Market Gardener by Eliot Coleman

Four Season Harvest: How to Harvest Fresh, Organic Vegetables from Your Home Garden All Year Long by Eliot Coleman

In a Pig's Eye by Karl Schwenke

The Vermont Papers: Recreating Democracy on a Human Scale by Frank Bryan and John McLaughry

Judevine: The Complete Poems by David Budbill

Blood Brook: A Naturalist's Home Ground by Ted Levin

Loving and Leaving the Good Life by Helen Nearing

American Intensive Gardening by Leandre and Gretchen Poisson

Orcas Island Library District